SHORT S[T

Piers Horner ha
  Cardiff Unive[
Civil Service for eight years before moving to
  Amsterdam, where he now lives.

Piers Horner

SHORT STORIES OF SPACE

For my grandparents.

CONTENTS

Acknowledgements

This book would not have been possible without the support of a critical band of people to whom I am indebted for their encouragement and feedback on early drafts. These include Jill Turnbull (and her science writer friend whose name I unfortunately do not know); David Hamson; Mark Wilcox; Rose Macfarlane, and Stephanie Browne. Final proofreading and editing were carried out by Sara Bigley and Jessie Raymond.

I am also extremely grateful to those academics who reviewed the more technical aspects of the stories. These include Professor Kanya Kusano (Institute for Space Earth Environmental Research, Nagoya University), Professor Edward Guinan (Villanova University), Professor James Schombert (University of Oregon), Doctor Luke Barnard (University of Reading), and the otherwise anonymous 'JK'. Any errors that remain, as well as any opinions expressed or inferred, are my own and have not been endorsed by any of the individuals or institutions listed above.

Even though it is said that you should not judge a book by its cover, Short Stories of Space would be lacking an essential part of its character without the stunning image that forms the cover art, which was very kindly provided by Wouter Verhesen. Thanks are also due to the assortment of people who provided feedback on the initial cover design via social media.

Last but by no means least, I would like to thank my fiancée, without whom I would never have started this book or had the resolve to complete it, and my parents, who picked through each story in the finest possible detail and were tirelessly patient in support of my interest in astronomy as a child.

Foreword

Space is one of those subjects that everyone tends to have an interest in. What is there not to like about the brilliance of the stars, the delicate beauty of nebulae, the raw power of a rocket launch, the many varied faces of the solar system's planets and their moons, the emotional drama of an eclipse, the mystery over whether life exists elsewhere, or the question of how the Universe evolved and why it exists at all? No matter what your preference, there is almost certainly something you will find fascinating about space.

My own obsession began as a young child and dominated nearly everything I did for the next twenty-something years. I joined my local astronomy club (the excellent Cardiff Astronomical Society) and probably developed a reputation as an insufferable know-it-all kid; I wrote letters to Patrick Moore and treasured every typewritten note he sent back, and I ended up going to University and studying astrophysics for nine years.

Then I hit a personal crisis of sorts and lost my direction. I left the field rather abruptly and ended up moving from South Wales to London to pursue a different path for a while. Unfortunately, however, nothing I did captured the same sense of amazement and wonder I had found previously in my studies of the Universe. It would be five more years, however, before I would hear my calling once more in the song of the stars, and I began rediscovering my interest in earnest.

One of the nice things about returning to a subject in which you once specialised is that you can also discover all those things you missed while your head was buried in your little corner of specific research. One memory, in particular, stands out - the time

I came across a remarkable movie of the infant star HR 8799. The time-lapse was made up of many different individual images taken over the course of seven years with the Keck telescope from the summit of the Mauna Kea volcano in Hawaii.

This movie was special because, in each frame, the light from HR 8799 had been removed, allowing the faint signals from the four planets that exist around it to be imaged directly. As if this were not extraordinary enough, the time-lapse allowed for their orbital movement to be detected. The resulting sequence shows planets beyond our solar system orbiting their parent star for the first time.

Watching these tiny blobs of light sweeping around their host is, for me, a remarkable experience. It is barely 30 years since the first planetary system beyond our solar system was detected, but the idea that we are already able to image such systems directly touched an emotion within me that I cannot quite put words to.

The movie was released in 2017 but, as it turned out, the first image of this planetary system was actually released in 2008 while I was still studying. Somehow, however, I had missed it at the time. In modern society, there is so much information available that it is impossible to keep in touch with everything, and it is easy to miss really exciting new discoveries or new perspectives on our cosmos. While high profile space events still make the headlines fairly frequently, there is a wealth of other news released every day that rarely get much widespread attention.

It is in the spirit of wanting to promote at least some of these lesser-known results that I first started writing Short Stories of Space. Over time, however, the remit of the work has broadened and evolved and now includes pieces that reflect on events that made headlines in the regular news media, such as the launch of SpaceX's Crew Dragon, as well as notable events, such as the thirtieth anniversary of the Hubble Space Telescope's service in space. Wherever possible, however, I have focused on stories that have seen significant research developments in 2020 or where I believe I can provide a different angle than that of the popular headlines.

There are, of course, many other stories that could have been written. Who can forget comet NEOWISE, which made its closest approach to the Sun in early July and became the brightest comet to grace Earthly skies for nine years? A month before NEOWISE reached its peak, scientists from the XENON1T Dark Matter Experiment - buried under the imposing face of the Gran Sasso massif in Italy - reported an unusual excess of events that could prove to be the first direct evidence for the existence a hypothetical particle called an axion, which could also be a component of the Universe's enigmatic dark matter.

Later in the year, astronomers appeared to have finally confirmed the source of mysterious events called fast radio bursts which had puzzled scientists since their discovery over a decade earlier, and the Japanese mission Hayabusa2 quite literally dropped off samples of material it had collected from the asteroid Ryugu before heading back out into space to explore new objects.

I am sure there are some who will say that a better choice of topics could have been made. In some cases, however, the omission is because I do not think the time has yet come to tell these stories properly. In others, it is because I did not have as strong a connection with the subject matter as I did with the tales that have ended up being included.

My only small regret is that I have not managed to tell the stories of some of the smaller events of the year, which were perhaps less significant in the grand scheme of things but no less magical. These included the first detection of extrasolar comets and a poignant tale of an interstellar planet that was discovered wandering lonely among the Milky Way, orphaned after being flung out of the system in which it was formed. Maybe I will be able to return to these subjects another day.

I should also note that a surprising (to me, at least) amount of debate has sprung up during the editing process about whether or not these short pieces really do qualify as stories. Clearly, in this I am biased, but I do nevertheless believe I am justified in calling them so. Hopefully, it is already clear that the subject matter is predominantly non-fiction, but stories have never needed to be fictional and I do not believe the title necessarily implies that they have been made up.

All the same I have attempted to incorporate the facts into a narrative style that is, for want of a better expression, more indulgent than would normally be used in an academic text or article, including regular flights of fantasy, and have sought to provide the reader with fairly clear narrative arcs.

Structure and stylistic character apart, there is also something I wanted to capture in each piece that alluded to the emotional or sentimental responses that the topics provoke. I am a firm believer that science cannot escape being a social activity. This is frequently perceived as a dirty accusation, as there is a sense that it undermines the central pillar of objectivity on which science is so fundamentally built. In my mind, however, the social nature of science is an essential and wonderful part of its essence. It matters that people find galaxies beautiful or that they get a feeling of being swept away when they contemplate the vastness, or the complexity, or the elegance, of the Universe as a whole.

As a result, I consider the experience of science to be as essential a part of its being as the equations of General Relativity, the complexity of biological systems, or the symmetries of the standard model. Science without wonder is a cold and empty shell, and it is an vital function of stories to speak to these emotions. If I have, in some small way, managed to do so in these Short Stories of Space, then I will be more than happy.

On the Right Shoulder of Orion

From as early as we can tell, humankind has ascribed patterns to the mighty canvas of stars that fill the night sky and used those patterns to paint the heavens with stories of legend and myth. In ancient times, it seems likely that knowledge of these patterns, which we now call constellations, would have been passed down between generations and would have been well recognised within the societies of the time.

As the centuries have passed and the creeping glow of streetlights have obscured the brilliance of the stars, our general knowledge of the constellations has declined. There are now only a relatively small number that are well recognised by the general public. One of the most familiar is the Great Bear, also known by its Latin title of Ursa Major, which means the same thing. The constellation can be seen throughout the year in the Northern hemisphere and is located high enough in the sky that it can generally be made out even in built-up areas. Its fame owes mainly to the easily recognisable seven bright stars that make the body and tail of the bear, although these are better known colloquially as 'the Plough', 'the Big Dipper' or 'the Saucepan'. In fact, the Saucepan only represents about a quarter of the entire arrangement of stars. The entirety of Ursa Major is much larger, covering about 3% of the entire night sky, making it the third largest constellation visible from either the Northern or Southern hemispheres.

When I was a child growing up, not too long after I first became fascinated by astronomy, I decided that I didn't mind Ursa Major. It was okay. There were some interesting objects in the region, for sure, but you needed a fairly decent telescope to be able to appreciate them properly. The stars were bright, but they didn't have the best sparkle except under the deep dark skies of the countryside. Perhaps worst of all, the pattern of the stars looked to me more like a stick version of a horse than a ferocious bear.

I had to wait until Winter began beckoning to see my favourite constellation. Given my love of Christmas, the long dark nights and the crisp cold skies, the wait only added to the magic. It was always a pivotal moment in the year when I first noticed the top half of the constellation sitting above the horizon or, creeping back to bed in the early hours of the morning, saw it in full, standing bold and proud in the sky, peerless among its neighbours, like a guardian of the night.

The constellation in question was Orion the hunter, the characteristic shape of which is perhaps only a little less well known than that of Ursa Major. Orion is one of the few constellations that requires only a marginal leap of imagination to see the form that it represents. Three stars across the centre of the constellation make up Orion's belt, below which is a cloudy region decorated with small stars that represents his sword. These are surrounded by four bright stars that make up Orion's shoulders and knees. A small group of three stars mark his neck or head.

In the great tapestry of the sky, Orion is surrounded by constellations representing his two hunting dogs, Canis Major and Canis Minor, and the small constellation of Lepus - an unfortunate hare being chased by Orion's dogs across the celestial sphere. The jewel in the eye of Canis Major - the so-called 'Dog Star', or 'Sirius' - is also the brightest star in the night sky, located below Orion's right knee, gazing up obediently at its master.

Not far from Orion's left shoulder is the dull red glow of the star Aldebaran, which forms the baleful eye of the constellation Taurus, the Bull. The horns of Taurus arch over Orion, threatening conflict and in my child's eyes the two were poised for action, frozen in the last moment of calm before battle.

Sadly, there is no link in the Greek mythology from which the story of Orion originates between the hunter and the bull. Little did I know then, however, that in earlier traditions of the Babylonian Epic of Gilgamesh, the 'Bull of Heaven' is sent to kill the eponymous Gilgamesh after he spurns the love of the goddess Ishtar. The seventh century Babylonians identified the pattern of stars that make up the constellation we now call Taurus with the Bull, and there is reason to believe they may also have identified the pattern of stars we now call Orion with the figure of Gilgamesh. It's the story my ten-year-old self would have preferred.

Another of Orion's attractions is the rich wealth of nebulae that fill the region and can be observed relatively easily. Nebulae are enormous clouds of gas and dust that form the birthplace of stars. Their dreamlike, complex beauty was one of the things that first enthused me about astronomy. The great Orion nebula that makes up Orion's sword is one of the most well-known and brightest examples of these in the night sky and can be seen relatively easily without needing a telescope. In residential areas, you might be able to glimpse a hazy form that is clearest when you aren't looking at it directly, like a ghost in the sky, but under crystalline skies the misty region stands out like a silver cloud, peppered with stars.

It takes a long exposure photograph to reveal the detailed structure of the Orion nebula: the twisted fingers of gas, sculpted by the winds of stars formed within or near it; the dark streams of dust that absorb visible light and the hollowed-out cavern at the centre of the nebula where a small cluster of newly born stars sparkle like a fistful of diamonds. The grand beauty of the Orion nebula is surely one of the most iconic images in astronomy, responsible for sparking the interest and imagination of many an amateur and professional alike.

Explore further and there are even more spectacular nebulae to be found. The Horsehead nebula is found at the right hip of Orion's belt and is formed by a distinctively shaped pillar of dust silhouetted against softly glowing hydrogen gas that rolls away like a gently unfolding landscape. Not far from the Horsehead is the Flaming Tree nebula, whose dark spine of dust, which forms the trunk of the tree, bifurcates to form its branches.

Zooming out and lengthening the exposure, one finds that the Great Orion nebula, the Horsehead and the Flaming Tree form part of an enormous network that covers the whole trunk of Orion. A great loop of gas, known as 'Barnard's Loop', cradles the three nebulae and stretches up to Orion's neck and head where another diffuse region, called the Lambda Orionis ring, forms a near perfect circle of emission.

It was, however, the stars that form Orion that first captured my imagination and perhaps form its most enduring appeal. Orion contains five of the top thirty-five brightest stars in the sky - more than any other constellation - and two stars in the top ten.

Coupled with nearby Sirius, the region stands out in the sky like a box of celestial gems.

Not only are the stars bright but they also have a distinctive set of colours. At first glance, most stars appear white to the naked eye. Allow your eyes time to adjust to the contrast with the dark background of the sky however, and you can sometimes tell that they have their own distinctive colours. Apart from the visual appeal, it turns out that astronomers can also tell a lot about the surface temperature, composition and age of a star based upon its colour.

The hues of stars are easiest to distinguish if you observe them through binoculars or telescopes. In some cases, however, the colours can be seen with the naked eye. The brightest star in Orion is one such example. Its name is Rigel, and it forms the hunter's left knee.

Rigel is a blue supergiant star located over 8,000 trillion kilometres from Earth. It is believed to be about twenty times more massive than the Sun and its surface temperature is around 12,000 Kelvin, making it tens of thousands of times more intrinsically luminous. To the unaided eye Rigel burns with an intense white, which occasionally flickers an electric blue.

It is on the opposite side of the constellation however - sitting on the right shoulder of Orion - that you find the constellation's second brightest star, glowing a distinctive rusty orange. This star is called Betelgeuse.

Betelgeuse is a unique object. Its mass is similar to Rigel's, yet its other physical characteristics are entirely different. Whereas the surface of Rigel is almost twice as hot as our Sun, the surface of Betelgeuse is relatively cool, at around 3,600 Kelvin. This difference in temperature is also the reason for their contrasting colours.

Betelgeuse is also one of the largest stars by diameter known in our home galaxy. It is so huge that if it replaced the Sun at the heart of our solar system, it would completely engulf the Earth. Its outer surface would probably extend well beyond the orbit of the asteroids, which lie between Mars and Jupiter, and the outermost edges of its atmosphere could even come within touching distance of Saturn. The extended atmosphere of Betelgeuse, which includes complex networks of gas and dust blown off its surface, is larger than our entire solar system. Its bloated body is so enormous that Betelgeuse is one of only a small number of stars in the night sky whose surface can be imaged directly by telescopes.

The physical differences between Betelgeuse and Rigel are mostly a result of their relative ages. Once a star is formed, it evolves over time and one day will die, either dramatically in an enormous explosion called a supernova or by gradually cooling into frozen oblivion. During its lifetime a star can undergo enormous changes in its size, composition and temperature which are driven primarily by the star's mass.

At a little under ten million years old, Betelgeuse has not existed for long compared to most stars. Our Sun, for example, is thought to have formed around four and a half billion years ago and will take the same amount of time again to exhaust its reserves of fuel before fading away. More massive stars run through their fuel much faster however, so the lifespan of stars like Betelgeuse and Rigel are much shorter.

Betelgeuse is slightly older than Rigel and represents a snapshot of what Rigel may look like in a few more million years. Toward the end of their lives, massive stars switch between different sources of internal fuel in an increasingly desperate struggle to produce enough energy to support themselves against gravitational collapse. As they do so, they swell up to enormous sizes, cooling and changing their colour as a result, to become a 'red giant' or 'red supergiant' star. This is the stage of its evolution that Betelgeuse has reached. One day, Betelgeuse will run out of fuel and lose its fight against gravity, and the subsequent rapid collapse of its dense iron core will trigger a supernova explosion, ripping apart the interior of the star and blowing it into space.

Betelgeuse is also a runaway; it is moving through space unusually fast, which has led to the formation of a large shock front that stretches far beyond the star itself as it ploughs through the diffuse material that lies between the stars of our Milky Way. The reason for Betelgeuse's unusual motion is unknown, as is the location of its birth. Many believe however that it originated in the great nebula of Orion itself, from where it was flung into space, perhaps as a result of another star's explosive death.

Whatever the cause, the unique character of Betelgeuse means it is well studied by amateur and professional astronomers alike. It did not take long therefore for the world to notice when, in November 2019, the star began to dim dramatically.

Villanova University in Radnor Township, Pennsylvania, is an institution dating back to the mid-nineteenth century. Founded by the order of Augustus in 1842, the soaring steeples of its church, the neatly manicured grounds and picture-perfect, tree lined avenues give the campus a fairy-tale feel. Richard Wasatonic and Edward Guinan are two members of the University's department of Astrophysics and Planetary science whose work includes pulsating stars and stellar evolution. For the past 25 years or so, Wasatonic and Guinan have meticulously documented changes in the brightness of a number of so-called 'variable stars' in the sky.

It is not unusual for evolved stars such as Betelgeuse to change their brightness. The stability of a star depends on a fine balance existing at different points of its interior between the buoyant hot gas - heated from below by the nuclear heart of the star's core - and the combined mass of the cooler layers that lie above it. Small changes to that balance can drive a cycle of expansion and contraction of the star, which leads to a variation in its brightness. There are a wide variety of variable stars which pulsate in this manner.

Changes in the apparent luminosity of a star can also be driven as a result of changes in its surface features. For most stars, the gas located in their upper layers churn and form great plumes of hot material that rise to the surface, release energy into the star's atmosphere, cool, then sink back into its interior. The tops of these 'convective currents' can be seen on the Sun's surface as a small-scale granular pattern, also known as 'convective cells.' Convective cells are small in comparison to the Sun's disc, but in other stars they can be gigantic - so much so that they can alter the star's apparent brightness as they first break through to the visible surface (leading to brightening), then cool and fade away.

The variation in brightness for many of these stars follows a regular, characteristic pattern that helps classify the exact mechanism driving the change. There are however some where the variation follows a less predictable pattern; these stars are referred to as 'semi-regular' or 'irregular' variables.

Betelgeuse is well known to be a semi-regular variable, so when Wasatonic, Guinan and Thomas Calderwood of the American Association of Variable Star Observers, first noticed a drop in the apparent brightness of the right shoulder of Orion, it was not immediately a surprise.

What they did not expect was that over the course of the next month and a half, the star would continue to dim quickly. On the 8th of December 2019, the trio of astronomers reported that Betelgeuse had reached its lowest apparent brightness since their records began. Astronomers around the world began to take note. As humanity entered a new decade, Betelgeuse slipped inexorably down the rankings of the brightest stars, losing its place among the top ten and even nudging out of the top twenty.

The question was, what was driving this dramatic level of dimming? No one could tell for sure, but one tantalising possibility was that Betelgeuse had reached the end of its life and was in the midst of its final turbulent death-throes before collapsing catastrophically to become a supernova.

If this were true, the Earth would have premier seats for one of nature's greatest and most terrifying spectacles. So much energy is released during a supernova explosion that they can, for a short period, outshine an entire galaxy of stars, making them easily visible even across the great gulfs that lie between galaxies.

As a result, even though supernovae are relatively rare by human standards (for a galaxy like our own Milky Way, we expect one supernova to happen every hundred years or so) they are relatively easy to spot for astronomers who regularly survey large numbers of galaxies to search for them. There are now networks of telescopes across the world that work together to follow up quickly on the detection of a new explosion, sometimes within minutes. This allows astronomers to gather as much information as possible at different wavelengths of light to help understand how supernovae explosions develop and evolve.

We have very little information, however, about what happens to a star immediately before it explodes. The closest supernova to Earth in recent history was in 1987 and was located in the biggest of the Milky Way's satellite galaxies, the Large Magellanic Cloud. The event became known, somewhat unromantically, by its assignment of SN1987A and, at its peak, could comfortably be seen from Earth without the aid of a telescope, despite being over 2,000 times further away from us than Betelgeuse.

The explosion of a well-studied and highly visible star in our galactic neighbourhood would, therefore, have been the cause of much excitement among astronomers and a truly spectacular sight. A supernova caused by Betelgeuse's demise would shine as bright as the half-formed Moon, visible in the sky by day and casting shadows on the Earth by night. Orion's right shoulder (or perhaps Gilgamesh's) would have been pierced by the Bull. His blood would have spilled into the night like the blossoming of a giant, bright flower. Less poetically, a detailed study of the star's behaviour in the lead up to the explosion would give scientists totally unique information about the physical processes involved.

No wonder, therefore, that once the popular press had picked up on the speculation, it became worldwide news. For weeks, the world held its collective breath as Betelgeuse continued to dim. The fainter it became, the greater the anticipation. Scientists smiled gently at the hype while secretly crossing their fingers and wishing for catastrophe; amateur astronomers trained their instruments at Orion in the hope of catching the moment that Betelgeuse would bloom, and members of the popular science media busily prepared obituaries for one of our most well-loved and evocative stars.

It was therefore a quiet disappointment to many when, as an unseasonably warm January rolled into an unseasonably warm February, Betelgeuse's brightness first stabilised then began to increase steadily once more. Over the remainder of February and March, it slowly recovered its brilliance and by midway through April it was even slightly brighter than it had been before the dimming began.

As it turned out, the recovery was almost completely consistent with predictions made by Wasatonic, Guinan and Calderwood when they alerted the world to Betelgeuse's behaviour. Semi-regular variable stars can have several 'modes' of brightening and dimming, each repeating over a different period of time. The astronomers had proposed that the unusual drop in brightness was a result of overlap between the minimum for two modes - one that normally lasts between about 5.6 and 6 years and the other of around 420 days. They had also predicted that the recovery in brightness would begin at the end of February. The case looked closed.

In a further twist however, images of the star's disc taken during its dimming phase by the pragmatically named Very Large Telescope (VLT), based in Chile, were compared to ones taken in January 2019 when Betelgeuse was shining at its normal brightness. It was immediately clear that there was something different. In January 2019, the star appeared as a relatively symmetric round disc, but in late December its lower half was much darker and looked as though it had been squashed.

In addition to the change in apparent shape of the star, researchers examined Betelgeuse's brightness at longer, lower energy wavelengths beyond the visible spectrum. Here too there was a surprise, for the data initially suggested Betelgeuse's temperature had remained consistent throughout the period.

The consistency of the star's temperature did not match with the idea that the change in brightness was due to overlapping modes. If the variation had been a result of Betelgeuse expanding and contracting, astronomers should also have seen a change in its temperature.

Suspicions as to the cause of Betelgeuse's strange asymmetric appearance and dimming began to turn in a different direction. Maybe dust was the culprit. It is well known that in later stages of their evolution giant stars blow material off their surface that can then aggregate in cooler environments farther away to form particles of dust. Dust absorbs visible light but is transparent at lower wavelengths of light.

Maybe Betelgeuse only appeared to dim in visible light because newly generated dust was obscuring the view from Earth. Once formed, the dust would gradually be dispersed by the continuous stream of particles emitted by Betelgeuse - its stellar wind - allowing the star to shine forth once more.

The question remained why the formation of the dust would align with the existing modes of variability such that the original prediction of Betelgeuse's recovery in late February would still come true. The coincidence in timing seemed unusual.

As the year progressed, the arguments have rumbled on. Observations by other groups have contradicted the idea that Betelgeuse maintained the same temperature during the episode; or have looked for evidence of outflows that could explain the formation of enough dust to reduce the star's apparent brightness but not found it. If these groups are correct, then the variation could have been due to a giant cooler region on the star, perhaps due to a gargantuan convective cell. By the end of the year, the jury remained out on what exactly had caused the unusual dimming.

In any case however, it seems as though Betelgeuse is no longer under immediate danger of exploding. Although we may have temporarily missed the opportunity to witness and study one of nature's greatest cataclysms in unprecedented detail, the ten-year-old me who spent hours outside transfixed by the stars would not have been quite so disappointed.

For while the supernova would have stood out in the night sky for several months and its afterglow would have been visible for years, within time it would inevitably have faded from sight. With it, Orion would have lost his right shoulder; the familiar pattern of stars would have been changed and the great herald of Winter's change would have been fundamentally disfigured. It is no exaggeration to say I would have felt as though I had lost a friend.

It is for this reason that when I now look to see Orion in the night sky it is with a sense of renewed appreciation, reminded as I am of the great privilege we enjoy of being able to see the many stars of the Universe in all their great wonder and beauty.

Where the Stars Shine the Clearest

I cons come in all shapes and forms, from the individual trailblazers who fundamentally change or help shape a particular field of human endeavour, to unique symbols, images or objects that have played some role in defining it. In the world of astrophysics, there are a handful of great instruments that have revolutionised humankind's view of the Universe and can rightfully be considered icons. There are few, however, that have been more beloved or more significant than the Hubble Space Telescope.

The space telescope, more commonly known simply as Hubble, was by no means the first instrument to be put in space, but it was the first large space telescope to take images of the Universe in visible light, as well as across a wide range of other wavelengths that are otherwise normally hidden from telescopes based on the ground. This versatility made Hubble an ideal tool for studying astronomical objects in a level of detail never achieved before.

As a result, the launch of Hubble in early 1990 opened an unprecedented window into the cosmos. April 2020 was the remarkable thirty-year anniversary of Hubble's operation, making it by far the longest operating telescope in space.

During those thirty years, Hubble has not just proven itself to be a cutting-edge instrument that has carried out ground-breaking research to advance humankind's knowledge of space. For many years, Hubble has been a household name as a result of its breath-taking images, which have probably done more to inspire new generations of astronomers and scientists than any other telescope or observatory. As a result, Hubble attained true legendary status among the scientific community and in the public domain.

The chronicle of the telescope's discoveries and its path to fame and success has not been without its dramas, of course. What is sometimes forgotten is that it is also the tale of the man who first set out in detail the dream of putting a large telescope like Hubble into space, and the woman who was central to making that dream become a reality.

The story of Hubble began almost forty-five years before it floated free of its clinical white cradle in the payload bay of the space shuttle Discovery. As the horrors of the Second World War wound to a close and the world began taking stock of the desolation that had been wrought, American physicist Lyman Spitzer was finishing his wartime research work and preparing to take up a new position setting up a new astrophysics group in Yale.

Spitzer had been a bright star of American physics prior to the war beginning, earning his PhD in Princeton under the supervision of Henry Norris Russell - who was an enormously influential figure in particular in the field of stellar evolution - before moving, first to Harvard to complete a postdoctoral fellowship, then to Yale as a teacher.

Although he initially opposed American intervention in the war, once the attack on Pearl Harbour forced the hand of Roosevelt and made US involvement inevitable, Spitzer became involved in what was then called Division 6 of the National Defence Research Committee, a section of which was devoted to undersea warfare research. In the final years of the war, he headed a group focussing on underwater sonar analysis and became involved in the work of the enigmatically named RAND Corporation.

Somewhat disappointingly as it turns out, RAND stood simply for 'Research and Development'. Set up as a spin-off from the Douglas Aircraft Company - which had contributed to the design and construction of the famous Flying Fortress air bomber - the aim of RAND was to maintain and strengthen the link between research and development and military planning, which proved critical during the war. (It would later become a not-for-profit research organisation claiming, among its other research credits, the development of packet switching - a method of bundling and distributing information that would eventually serve as the technological basis for the internet.)

Spitzer himself characterised RAND as more of a study group aiming at exploring a range of novel research areas. As a self-confessed avid reader of science fiction in his earlier years, the opportunity to indulge in blue skies thinking must have been particularly attractive to him. One of their areas of interest was the potential applications of a so-called 'world-circling spaceship' and in 1946, Spitzer wrote an internal paper that discussed the scientific opportunities that would arise as a result of putting a telescope in space.

At the time however, the paper was not published and as rocketry was a field still very much in its infancy, the practical problems with locating and operating a telescope in space meant that it was not seriously considered by the wider scientific establishment. Much early work on attempting to fit scientific instruments to rockets was carried out by small groups of enthusiasts while the rest of the research community focused largely on balloon borne experiments.

Twelve years later in 1957 however, the Soviet Union became the first nation to successfully launch an artificial satellite into orbit around the Earth. The satellite, called Sputnik, came as something of a shock to the West and in particular to the American public, where anti-Communist rhetoric had worked hard to portray the Soviet Union as a technologically backward state. Here was incontrovertible evidence that - in terms of developing a presence in space at least - they were in reality some way ahead.

The situation got worse as American attempts to launch its own satellite just over a month later ended in miserable failure as the rocket barely lifted off the ground before stalling and exploding as it flopped back down onto the launchpad.

Western nations were gripped with the fear that if the Soviets gained the ability to operate in space, they could pose a strategic threat against which democratic nations would be helpless. Insensible to the consternation its presence was causing, brave little Sputnik nevertheless embodied that fear perfectly - the regular radio beeps it emitted as it flew overhead delivering a message that could neither be blocked nor ignored.

America redoubled its space programme and a year later set up the National Aeronautics and Space Administration, better known as 'NASA'. Despite the military tension, NASA's remit was strictly civilian in nature and aimed to rebuild the United States' reputation for space exploration. The foundation of

NASA also boosted American efforts to put telescopes into space. In 1962, it launched the first Orbiting Solar Observatory (OSO), which was designed primarily to collect data about the Sun at high energy wavelengths.

Similar to the Russian Proton satellites, which were sent into orbit a few years later and were intended to study cosmic rays, the OSO missions were more designed to be space-based platforms for experiments rather than telescopes in the sense that they would be popularly recognised today. In a technical sense however, the basic idea of a space-based telescope had started to take form.

While all this had been going on, Spitzer had taken over as director of Princeton University Observatory and Chair of the Astrophysical Sciences department from his former supervisor Russell. He had already established a strong reputation for his work on the formation of stars and swiftly become a founding figure in the study of the diffuse material located between the stars of galaxies, also known as the 'interstellar medium'.

Spitzer had also carried out significant work in the early study of plasma physics, which deals with the behaviour of hot gases composed of bare nuclei and free electrons. In the early 1950s, he managed to convince the US Atomic Energy Commission to support early research into nuclear fusion. This led to the establishment of the Princeton Plasma Physics Laboratory which he also headed for many years.

Spitzer was keeping busy, but he had never forgotten the dream of a space telescope and by the end of the 1950s, he was no longer alone.

In 1960, Nancy Grace Roman had taken up the post of Chief of Astronomy in the Office of Space Science at NASA and became the agency's first female executive. In a profession and society still largely dominated by men, Roman had needed to demonstrate resolute determination to pursue a career in science from an early age. The direction given to Roman by her high school counsellor was that women simply could not become astronomers and she was encouraged to study Latin as opposed to mathematics as a subject more befitting her gender.

Happily, Roman paid no attention whatsoever to this advice and quickly proved herself both talented and tenacious, despite the litany of discrimination she faced during her academic career. Eventually, Roman found her way into Government work which she found markedly less discriminatory than being in the University system.

Even before she joined NASA, Roman was already seemingly interested in the prospect of space-based instruments. She worked on the OSO mission and quickly became involved in their successor programme - the Orbiting Astronomical Observatories (OAO) - after joining NASA in early 1959.

Still, the proposed design for the OAO satellites were quite small in comparison to the much bigger space telescope Spitzer had envisioned. In 1962 however, the National Academy of Science convened a conference at the University of Iowa to provide advice to NASA on a range of areas. Among the discussions that took place, it was pointed out that rockets were becoming powerful enough to launch a large telescope into space, which created a lot of excited discussion about the feasibility of such an instrument. There were however technical challenges being able to accurately point a remotely operated telescope floating in space at faint objects. Given her experience with the OAO programme, Roman was unconvinced that such a large, complex project was deliverable at the time.

The topic was returned to in a second study group held in Woods Hole in 1965. By this stage there was much more confidence among academics that the project was possible, and the meeting ended up issuing advice to NASA that a large telescope in space would become an increasingly important tool in the advancement of astronomy.

Roman continued to believe that the timing was not yet right, but the idea had started to develop a life of its own. A number of companies had begun developing proposals for manually operated telescopes in space, but these did not match the needs of the astronomical community. Somewhat reluctantly, it would seem, Roman decided that matters needed to be taken in hand. She also realised, however, that one of the first tasks that needed to be addressed was ensuring the rest of the scientific community was also on board with the idea.

With the benefit of hindsight, it is hard to imagine that anyone would have opposed plans for a space telescope, least of all members of the community it would so end up serving so faithfully, but the project came with a price tag that reflected its ambition. The cost of design, development, construction and testing of the telescope would be huge and given that the instrument needed to be transported into space, the risk of total failure of the mission was a real one. If anything went wrong during launch, for example, all that money, time and effort would have been wasted.

It is a common dilemma in the funding of space projects. In addition, however, the space telescope would need to be maintained while in space in order to extend its lifespan as much as possible, meaning that the cost of servicing missions during its operational lifetime also needed to be considered. With this in mind, it is perhaps less surprising that there were still many who believed the investment was better spent supporting much cheaper development of ground-based observatories and their facilities.

Shortly after the Woods Hole meeting, therefore, a working group was set up to establish the goals of a space telescope (which also became known at the time as the Large Space Telescope, or

LST), with Spitzer as the Chair. Over the next few years, the work of the committee as well as the dedicated advocacy of Spitzer and Roman helped to begin turning the tide of academic resistance. In 1968, NASA began drawing up plans to take forward the project, and in 1971, Roman established a Science Steering Group to work on plans that would eventually become the blueprint for Hubble.

The LST had finally taken its first step off from the pages of Spitzer's 1946 paper. As with any major project however, the progression from concept to planning was only the first step in a long and difficult path, which would so often be crossed by many formidable barriers.

While their internal lobbying had shifted the scientific community in favour of the LST, a much greater challenge lay ahead in convincing the grey suits of congress who would control the flow of state funding for the project. The optimism of the post-war era had worn off and amid the ongoing threat of the Cold War, the world was entering a period of uncertainty accompanied by cuts to public spending.

Roman and other NASA administrators worked hard to build support among political figures, hosting dinners at which Roman would deliver talks about the science goals of the LST and providing presentations to the Bureau of the Budget, which was responsible for preparing the budget for the President. It is largely for her pivotal effort in the struggle to secure both academic and political backing for the project that Roman is remembered as the 'Mother of Hubble'.

Still, however, things remained on a knife-edge. The year 1973 saw the start of the biggest recession to affect Western nations since the Great Depression of the 1930s. Times were hard, even for the biggest economy in the world and Capitol Hill demanded that the ambitious plans for the LST be scaled back. In 1974, during the worst period of the stock market crash, LST funding for the following year was cut completely. For a time, the project looked to be dead.

It was at this crucial moment however, that Spitzer and Roman's earlier advocacy for the LST proved its worth. A national lobbying campaign sprang up supported by the National Academy of Sciences. The concept clung on to life. Eventually, the recession began to ease and funding for the project was reinstated. In 1977, congress gave its approval for the project to go ahead.

Work on the LST could now begin in earnest. A launch date was set for 1983, but it would be over a decade until the telescope was ready as issues in the production of its main mirror - which needed to be ground to an extremely precise shape to optimise its performance - led to delays.

Still, 1983 proved a year of symbolic importance to the LST as the somewhat unsentimental moniker was dropped and it was re-named the Hubble Space telescope, after the American astronomer Edwin Hubble. The statement of intent was clear. Hubble's work in the 1920s had revolutionised humankind's view of the Universe by demonstrating that some of the oddly shaped nebulae astronomers observed in the sky were actually independent galaxies that lay beyond our own Milky Way, and by providing the first evidence that the Universe is expanding. Just as Hubble had helped rewrite our understanding of the cosmos, so too would the observations from the Hubble Space Telescope.

As the year 1986 dawned, Hubble looked set to launch in the Autumn. Five years prior NASA had carried out the first successful mission of its space shuttle fleet and Hubble was due to be carried into space locked into the belly of one of the flagship spacecraft.

Then, tragedy struck. Slightly over a minute after the space shuttle Challenger had lifted off from Cape Canaveral on the 28th of January 1986, the rockets to which it was attached disintegrated and the spacecraft broke apart at an altitude of around 14 kilometres. All seven crew on board were killed, among them a schoolteacher from New Hampshire, Christa McAuliffe, whose inclusion in the team had led to widespread public interest in the launch. Spectators on the ground watched in horror as the spacecraft was enveloped in a cloud of vapour and replaced by a myriad of fragments, arching through the sky like streams of ticker tape at a parade.

The publicity around the Challenger launch had taken a heart-breaking twist. The entire space shuttle fleet was grounded as Ronald Reagan, who was due to deliver his State of the Union address on the same day but delayed it because of the accident, launched a commission to investigate what had gone so catastrophically wrong. America's space programme was on hold and Hubble was locked away, all ready to go yet unable to make it into space.

It would be four long years following the Challenger disaster before Hubble was unpacked once more and carefully transferred into the protective interior of the space shuttle Discovery. Finally, in April 1990, the shuttle and its precious cargo - upon which so much expectation was resting - were at last rolled out onto the hot tarmac of the launch pad at the Kennedy Space Centre, ready to begin their ascent into the sky.

Hubble was by no means the largest of instruments. Bragging rights for telescopes are typically based on the size of their primary optics - the main mirror or lens that focuses light entering the telescope - which also defines the smallest theoretical scale that can be resolved in the night sky, known as the telescope's angular resolution.

Hubble's main mirror was 2.4 metres in diameter and while this was not to be sniffed at, it was much smaller than the 6-metre mirror of the world's largest telescope at the time of launch, the

BTA-6 in Russia. By 1993, the BTA-6 had been overtaken by the first of the mighty Keck telescopes, located on the peak of the dormant Mauna Kea volcano in Hawaii, whose primary mirror spanned a whopping 10 metres.

What Hubble lost out on in size, however, it more than made up for in location. When a ground-based telescope observes the night sky in any given direction, it has to do so by looking through the atmosphere. Currents in the atmosphere constantly blow areas of higher and lower density air across the line of sight of the telescope, each of which bend the light from whatever source is being observing in slightly different ways, distorting the image very slightly. As a result, over the duration of an observation, the light from a point source such as a distant star, or the detail of an extended source such as a galaxy, becomes smeared out over a larger area.

This effect is called atmospheric seeing and limits the resolution that can be achieved by a classical telescope regardless of its size. This is one of the reasons that many telescope observatories are located at high altitude, to avoid observing through the lower, more turbulent layers of the Earth's atmosphere.

Sat at around 540 km above the Earth's surface, however, Hubble would be located above the vast majority of the atmosphere, so would avoid having its resolution downgraded as a result of seeing. Despite its modest size, Hubble was expected to produce images that were sharper and revealed more detail about the Universe than was ever possible with even the biggest ground-based instruments.

Hubble had another crucial advantage too. The molecules that make up our atmosphere, in particular ozone, water and carbon dioxide, absorb light at certain energies, blocking it from reaching the Earth's surface. In the case of ultraviolet light, which has a higher energy than visible and causes damage to biological cells, this absorption is essential for life on Earth. Life's gain is, however, a problem for observers. From the ground, astronomers can only observe the Universe using light from a limited range of wavelengths called atmospheric windows. At other wavelengths, it is as though the Earth were surrounded by a dense cloud layer that blocks out all signals from space.

Above the atmosphere there is no such problem. The Copernicus mission that Lyman Spitzer had helped to design had done much to trailblaze a more detailed understanding of how the Universe looked at ultraviolet wavelengths and X-rays. Now Hubble would go one step further, with a more powerful set of optics and instruments that could observe the Universe not only in ultraviolet and visible light, but also broad regions of the infra-red spectrum that were otherwise invisible from the ground.

In short, Hubble promised to be the gateway to a whole new perspective on the Universe. It was this vision that Spitzer had embraced and incubated in his 1946 paper, and which made the decades-long effort to see it hatch and grow worth the investment. Scientists around the world watched and waited for Hubble to open its eyes.

The first picture from Hubble was released a little over three weeks after the launch. For such a landmark moment, the image itself was markedly understated - just a humble field containing a small selection of stars used to begin fine-tuning the focus of the telescope - but the superiority of Hubble's resolution compared to ground based instruments was already evident. It looked as though the space telescope would indeed live up to the hype.

It was not long, however, before technicians noticed a problem. Hubble's performance was good, but not as good as it should be. Extended celestial objects that should have been pin-sharp instead appeared cloudy, as if they had been smudged out slightly, and bright point-like sources were surrounded by feathery patterns of light that reduced considerably the sharpness of the image.

The source of the problem was quickly identified. The troublesome manufacture of the primary mirror had not only delayed the launch of Hubble; there had also been a mistake made in the shaping of the mirror's surface. The margin of error was tiny: a wayward millimetre of additional separation between two components of the lab instrument used to test the mirror's shape while it was on the ground was to blame for its outer edges ending up two thousandths of a millimetre flatter than they should have been. If the design had been correct, all the light rays reflecting off the mirror would have converged to the same point or focus. As it was however the focus depended ever so slightly on which part of the mirror incoming light reflected off.

Small though the mistake was, it made a significant difference to Hubble's ability to resolve faint objects, in particular in the visible and infra-red portions of the spectrum. Luckily, the imperfection of the mirror could be compensated for by inserting the right combination of specially shaped lenses and mirrors into the path of light as it passed through the instrument.

The obvious issue was that Hubble was in space. Scientists would have to wait until the first servicing mission in December 1993 for the opportunity to correct the faulty optics. Even then, the mission was full of risk. NASA had planned no less than five spacewalks over eleven days - more than on any previous mission - to maintain and fix Hubble. The seven members of the mission's crew were among the Agency's most experienced astronauts and had undergone almost a year of intensive training for the flight, involving the use of hundreds of bespoke tools.

The Administration's reputation was on the line once more. With memories of the Challenger disaster still casting a shadow over the ambitions of the manned spaceflight programme and only mixed success with recent satellite projects, failure of the servicing mission would not only undermine Hubble's science objectives but also cast doubt on NASA's capacity to deliver major projects, including the upcoming International Space Station.

All of which must have added an enormous burden of expectation to the crew during the mission, as if counteracting the apparent weightlessness of being in space. Suspended above the shimmering blue oceans and dusty brown continents passing by hundreds of kilometres below and guided by an astronaut tethered to the end of an insect-like white robotic arm, the housing for a set of corrective optics and a replacement for one of Hubble's main detectors was manoeuvred slowly into place.

The fix worked. In a landmark press conference in January 1994, NASA officials, the White House director of Science and Technology policy, and Maryland Senator Barbara Mikulski announced that the "trouble with Hubble" was over. Before and after images taken with the telescope showed a significant improvement, demonstrating that Hubble's sight was sharp and clear.

Almost fifty years after Spitzer's paper, and after having come so close to the brink of failure and abandonment, the dream of the large space telescope was finally being realised.

Over the many years since, there is barely an area of astrophysics that Hubble has not influenced in some way. From seminal work confirming the existence of black holes to advancing our understanding of the physical processes involved in the birth of star and the formation of planetary systems and our measurements of distances to galaxies, Hubble has brought new insights that have moved astrophysics forward at a rate that would have been impossible without it.

Along with the science, Hubble has produced images that have become almost as iconic as the telescope itself. I still remember as a teenager hearing on the evening news about the release of an image of columns of gas and dust several light years long in the central region of the Eagle nebula, evocatively named the 'Pillars of Creation'. Apart from being a spectacular picture, the appeal of the image was the story it told of the dynamic processes of star formation that take place in the heart of stellar nurseries. From the glowing ends of the pillars themselves, sculpted by the ultraviolet radiation of nearby young stars, to the tadpole-like globules of gas where new stars are forming. Here, before our very eyes, stars and planets were being born and Hubble gave us a ringside seat to watch the spectacular action.

Hubble has also borne witness to drama located much closer to home, in particular the spectacular collision between the fragments of comet Shoemaker-Levy 9 and Jupiter in June 1994. Hubble took defining images not only the fragments themselves, strung out like beads as they hurtled along the final stages of their approach to the giant planet, but the dark brown scars that pocketed Jupiter's face long after the impact.

In my mind however, nothing quite beats a composite image taken by Hubble of one of the darkest, emptiest patches of sky that the project's scientists could find. Over a period of ten days straddling Christmas 1995, Hubble stared for over one hundred hours at a single patch of apparently empty sky, chosen for being almost entirely devoid of objects from either our Milky Way or the nearby galactic environment.

The results were remarkable. This obscure patch of nothingness was in fact teeming with galaxies of all shapes and sizes. The Hubble Deep Field, as the image came to be known, contained over 1,500 galaxies spanning billions of years of the Universe's history and including some of the furthest objects known at the time. An entire history of galaxy formation was laid out in a single view, from the tiny red smudges of light that barely showed up on even this deepest of images, but which represented the first building blocks of galaxies, to swarms of distinctively shaped galaxies of all types.

The scope of the Hubble Deep Field was quite literally breathtaking and even though it has since been superseded by later Deep Field images taken with Hubble, it is still surely one of the most significant pictures in astronomy.

There is a particular quality to Hubble's images that makes the viewer feel as though they are part of the scene, whether falling into the midst of a giant galaxy; or surrounded by brilliant swarms of stars in a globular cluster; or floating in space not far from Jupiter or Saturn. No wonder, therefore, that Hubble is a popular and cultural ambassador for science like no other.

If there is a bittersweet note to the story of Hubble's success it is that the telescope has outlived both of its honorific human parents. On 31 March 1997, Lyman Spitzer collapsed at home after an otherwise normal day in the leafy surroundings of Princeton University. Fittingly, the work he was pursuing at the time was using data from Hubble. At the end of 2003, following the results of a public essay contest, NASA named the final instrument of its Great Observatories space programme, which had begun with Hubble, after him.

In late 2018, Nancy Roman passed away in hospital. Her fight for Hubble had been won a long time ago but Roman had continued to popularise astronomy, in particular to women and girls, throughout the later years of her life. In 2020, NASA announced that they would also name their next generation flagship infrared space telescope after her.

The poetic symbolism is hard to ignore. First Spitzer, then Roman, taking their well-earned place out in the inky blackness of space alongside their prodigy the Hubble Space Telescope, in the place where the stars shine the clearest.

The
Unpredictable
Constant

E very field of science has its rockstar disciplines - those areas of study in which even the most mundane update is guaranteed to draw attention and publicity, and which everyone aspires to study and understand. In the case of astrophysics, cosmology is the undisputed bearer of one of its rockstar crowns.

One of the reasons for cosmology's appeal, of course, is that it more readily touches a deeper and more existential nerve. What question in science is grander or more mysterious than that of why the Universe exists at all? There is a debate to be had, of course, about whether science overreaches when it tries to lay claim to the right to answer that question or whether it is best left to philosophy or religion to sort. Despite its worthy fascination, the debate has been played to stalemate countless times in the past, although it continues to be repeated online every day through long comment threads in which everyone ends up convinced they have won.

Whatever your view on which branch of human learning is the least inappropriate for answering the ultimate question of origins, it seems clear not only that science has the right to try, but that it also provides some fundamental insights that demand consideration. The discovery that the Universe appears to have had a beginning, for example, offered a profound statement about the nature of reality itself that was, in a way, more specific than its alternative. It said that the existence of our Universe could not simply be taken for granted. It lent backing to the notion that the manifest of the cosmos, within which we define all that is physically real, may itself be transient in nature rather than immutable and eternal.

Even without the question of ultimate origin, however, cosmology is undeniably fascinating. What definitely does lie within the grasp of science is a description of how the early Universe looked, behaved and evolved, and how it transformed into the one we see today. As the phrasing suggests, this task involves attempting to grapple with the notion of a cosmos that is entirely different to anything we have experience of.

Not only was this early cosmos one in which stars, galaxies and planets simply did not exist, but it contained forms of matter more extreme than can be found anywhere in the current Universe and fundamental particles that no longer exist today. Even space and time are believed to have behaved in ways which are completely unfamiliar and counterintuitive to us now. The story told is mind boggling, both in its scope and its bewildering exoticness. Cosmology does not need the origin question to be cool.

Modern cosmology began during the interwar period in 1929 when Edwin Hubble published results that demonstrated a relationship between the distance of galaxies and their apparent speed of recession from the Earth.

It is an oft-repeated fact that when we look into the night sky we are also looking back in time. This is not a result of some distortion of the dimension of time but is a straightforward consequence of the fact that light travels at a finite speed through space. As a result, the further away an object is, the longer its light takes to reach the Earth or, equivalently, the earlier it has to leave the object in order to reach us at the same time as light from a closer source.

To make life easier for themselves and everyone else, astronomers define one of the fundamental scales for measuring distance in the Universe to be the 'light year', which is the distance travelled by light through an empty vacuum in one year. Light emitted from a star located at a distance of fifty light years from Earth will, therefore, take fifty years to reach us, and when we observe that star, we are viewing it as it was fifty years ago. Likewise, light from a galaxy located two million light years from Earth will have taken two million years to reach us, and we will be seeing that galaxy as it was around the time that the first recognisable ancestors of modern-day humans began walking the Earth.

This link between the distance of an object and the time at which its light was emitted meant that the relationship Hubble had uncovered between the distance of galaxies and their recessional velocities was actually revealing something about the expansion history of the Universe. This was because the recessional velocities measured for galaxies are interpreted as being a result of the expansion of space itself rather than their own individual motion.

By measuring recessional velocities at a range of distances therefore, Hubble was effectively sampling the rate of expansion of space at different points in history. His observations appeared to demonstrate that space was expanding faster in the past and that the speed of expansion increased consistently the further back in time you observed.

This was the fundamental observation for which Hubble became most famous, although for a long time Hubble himself expressed reservations about whether or not the measurements really did measure galactic recessional velocity, as opposed to some other aspect of the galaxy or the light it emitted. Today however the methodology is widely accepted, and Hubble's measurements form the cornerstone evidence that supports our understanding that the Universe is expanding. The apparent constant rate of increase in expansion velocity could also be calculated and instantly became a value of fundamental importance in describing the Universe's evolution, known as 'Hubble's constant'.

Getting a constant named after you is one of the ultimate achievements of a scientific career. In humankind's perpetual quest to secure immortality through history, being associated with a fundamental property of nature seems like a far more durable and less personally risky way of preserving one's name than, for example, achieving bloody victory in battle. It may therefore have come as a source of very marginal disappointment to Hubble that his constant was quickly shown to not be a constant at all.

As we now understand, the expansion rate of space changes depending on what form of energy dominates in the Universe. As a result, Hubble's constant is better thought of as the Hubble parameter, whose value changes depending on what epoch of the Universe's history is being examined. During its earliest phases, most of the Universe's energy density came from light and other massless (or near massless) particles. The energy density of radiation decreases relatively quickly as the Universe expands however, and at some point, the energy density of matter - which reduces less quickly - began to dominate instead. This shift in the relative importance of matter and radiation also led to a change in the expansion rate.

For much of its early and middle ages, the Universe remained in a largely matter-dominated regime which, to be fair, kept the expansion rate reasonably consistent. Nevertheless, it has continued to change, and we now know that over the past few billion years the cosmos has even been accelerating. Hubble's observations were constrained as a result of him only being able to take measurements of galaxies within a limited range of distances (or equivalently, a limited period of time in the Universe's history), over which changes in the expansion rate were too small to be observed.

The accuracy of his results was also limited as a result of uncertainties in some of the underlying principles and relationships that allowed him to measure the distance to the galaxies he studied. Regardless, Hubble's work opened the gateway for cosmology to flourish and he is rightly immortalised as one of the field's founding fathers.

In the decades to come, one of the main challenges for cosmology was to understand the implications of Hubble's discoveries for the early history of the Universe. If the expansion was extrapolated backwards in time without limit, the diameter of the Universe would eventually shrink to zero. This point-like source would presumably represent the initial state of the Universe, out of which everything subsequently evolved. So it was that the Big Bang theory was born.

The philosophical implications of the Big Bang hypothesis were not popular with everyone. The concept of a Universe that was came into existence in a single moment of time seemed uncomfortably reminiscent of the creation myths of religious teachings. An alternative proposal to explain the Universe's behaviour emerged that claimed matter was continuously being created in the emptiness of the vacuum, and that this emerging matter drove the expansion of space in a Universe that was otherwise infinite and had existed forever. This proposal became known as the Steady-State theory.

The ensuing academic tussle between the Big Bang and Steady-State theories continued until 1965, when the discovery of a uniform background of millimetre wavelength radiation coming from every direction in space provided evidence that the Universe had once been much smaller and hotter. Over time it became clear that the Steady-State theory simply could not explain the precise way in which the energy of this background was distributed, and by the early 1970s it had lost favour, leaving the Big Bang theory to reign supreme.

While the triumph of the Big Bang theory provided clarity on the early state of the Universe, however, there were some issues that suddenly became harder to solve as a result. A particularly thorny example was the so-called 'horizon problem', which derived from the deceptively simple observation that the Universe appeared virtually identical if you looked in one direction of the night sky as it did if you looked in the opposite direction.

The reason this was difficult to explain is that physical systems tend to only end up looking the same everywhere if there has been plenty of time for different parts of the system to mix. Milk poured into a mug of hot water that also contains a tea bag takes time (and usually a few brisk stirs of a teaspoon) to reach the optimal, uniform brownish colouration that promises a consistent and satisfying flavour. Similarly, if you find that the relaxing bath you are drawing has ended up too cold for comfort and adjust the taps so that only hot water is being added, it will take time for the energy from the hot water to become distributed across the entire volume of water.

In order for a physical system to reach a uniform state, there needs to be some means of transmitting information about the differences in its physical state throughout it. In the case of the bath, the thermal energy of water from the hot tap is spread out through conduction between molecules. Warm water atoms vibrate more vigorously than cold ones and some of this extra vibration gets passed on, increasing the vibration of surrounding molecules, which in turn affect the molecules around them, and so on. As the vibrations spread, this is the same as saying that the heat gets distributed throughout the bath.

Crucially, you can also think of this process in terms of an exchange of 'knowledge' about the extra energy of the hot water. In a sense, the vibration of the water molecules acts like a communication network passing on the message that 'hot water is entering the bath'.

The same basic principles about the brown drink and the bath of water should also hold true for the cosmos as a whole. It is extremely unusual for a natural system to assemble in a perfectly uniform condition. As a result, the standard assumption in cosmology is that the Universe would also have begun in a state that was not uniform, and that any differences between regions would only have been smoothed out via the exchange of information between them.

This is where the horizon problem came in. Einstein's theory of relativity tells us that it is impossible to accelerate an object to a velocity that is faster than the speed of light in a vacuum. The speed of light therefore represents a fundamental limit on how quickly information can be transmitted between two locations and, as a result, on how quickly a non-uniform system can transition toward uniformity. The issue was that when looking as far back as we can in two opposite directions in the night sky, the light we detect has only just had enough time to reach us. If the light - coming from opposite ends of the Universe - has only just made it to Earth, then there seems to be no way in which any information could have been exchanged between these two regions of space.

If that logic is correct, however, then why did the regions look the same? Either the Universe began that way - which begged the question of why - or else somehow the regions must have found a way around the communication limit imposed by relativity and have exchanged information with one another by some other route.

An answer to the horizon problem eventually began to take shape in 1979 when Alan Guth stumbled upon the mechanism of inflation. At the time, Guth was not trying to solve the horizon problem. Instead, he was interested in the question of why we do not see any isolated magnetic North or South poles in the cosmos - so-called 'magnetic monopoles' - even though they were predicted to be generated in large numbers by various theories of the early Universe.

Guth realised that the presence of a new particle, identified with a specific type of mathematical entity known as a 'scalar field', could, under the right conditions, drive the expansion of the Universe at a rate that increased exponentially. What was more, because the expansion was of the dimensions of space and time themselves, it was argued that the maximum expansion speed that could be reached in this way was not limited by the usual laws.

For Guth's purposes, this astonishing expansion would serve to separate magnetic monopoles to such an extent that they would become very rare in the bit of the overall Universe we ended up in. It also however offered a potential fix to the horizon problem.

If it so happened that, early in the Universe's history, a scalar field existed that produced a period of exponential expansion in the dimensions of space and time, this could have the effect of separating two regions of the Universe at a rate that was much faster than the speed of light. If these regions were in contact prior to the start of the exponential expansion, then there may have been enough time to smooth out any differences between them before they were swept apart and lost contact.

This is the concept that lies behind the theory of inflation, which has since become an essential element in our modern understanding of cosmology. The story goes that inflation happened in the barest moment of time after the start of the Universe and lasted for less time than it takes for light to cross an atom. Yet in that most fleeting of moments the dimensions of space increased their size by a staggering amount, flinging apart different parts of the Universe to such an extent that even now, many billions of years after the beginning, light has still not managed to catch up.

It was not long before scientists realised that the supercharged expansion of inflation solved a number of other conceptual problems with the Big Bang model as well. One of the most important for providing a bridge to the current-day Universe was answering the question of how the structures we see in our cosmos formed.

Most people are familiar with galaxies as one of the basic units of matter on large scales, but galaxies are themselves found to be grouped into clusters, and the clusters into enormous super-clusters spanning hundreds of millions of light years. Together, the entire ensemble of galaxies, clusters and superclusters form a lattice-like structure that threads its way throughout space, with superclusters marking the edges of gargantuan voids where very little structure is found.

At scales such as these, the only force significant enough to drive major changes in the way matter is organised and arranged is gravity. The standard view, therefore, is that all the structure we see today must have originated in variations in the distribution of matter early in the Universe's history. Over time, these 'seeds' of large-scale structure would grow, as regions that were slightly denser than their surroundings began pulling in more matter through their gravitational influence and would eventually go on to form the entire hierarchy of structures we see.

At first sight however, inflation would seem to be exactly the wrong mechanism for nurturing this kind of process. The rapid expansion needed to guarantee uniformity across the visible Universe - thereby solving the horizon problem - should have also completely smoothed out any irregularities in the infant Universe prior to inflation that could have acted as the seeds for structure later on.

Inflation's saving grace comes in the unexpected form of small random fluctuations that occur on the smallest scales in nature, deep within the realm of quantum physics.

These quantum fluctuations are a result of an inherent uncertainty that exists in nature when we try to measure properties of space and time at these tiny levels. Scientists realised that inflation would stretch these miniscule variations up to much larger scales, amplifying the imprint of quantum uncertainty onto all the matter in the cosmos to produce the necessary seeds from which structure could form.

The idea was as outrageous as it was elegant. By the time the Cosmic Microwave Background Explorer satellite - better known as COBE - was launched in 1989, scientists had also worked out that these fluctuations in matter would leave an imprint on the background of radiation discovered by Penzias and Wilson. Simply put, over-dense regions of the Universe would be slightly hotter and under-dense regions slightly cooler. This would lead to slight differences in the energy of the background light coming from the direction of these hot and cold spots called 'anisotropies'.

One of COBE's primary objectives therefore was to map Penzias and Wilson's background - more commonly referred to as the cosmic microwave background, or CMB - across the entire sky and try to detect these anisotropies. The task was not easy: the expected level of variation was only one part in a hundred thousand, so the difference in signal due to the over and under-densities was tiny. In 1992 however, the COBE team triumphantly released results from the first two years of observations that clearly showed anisotropies in their maps that closely matched the predicted values.

Cosmology was moving from strength to strength. Having established the key elements of a model describing the early phases of the Universe's evolution, one of its central questions in the lead up to the end of the twenty-first century was what the ultimate fate of the cosmos may be. To begin with there were, broadly speaking, only two possibilities, neither of which seemed particularly attractive.

The first option was that the Universe would continue expanding forever, cooling all the while as stars burned themselves out and galaxies gradually faded. Eventually the Universe would become a dark, frozen wasteland devoid of form or light, the only saving grace of which would be the occasional explosion of an evaporating black hole and the potential decay of protons.

The other option was that the Universe would eventually stop expanding and start contracting once again, presumably collapsing once more into a single point indistinguishable to that from which it was originally formed. At least here there was a sliver of hope. The idea of the Universe returning to its initial state offered the possibility that the collapse could itself be followed by a new Big Bang, which would recycle and renew all the constituents of the cosmos, wiping the slate clean ready for its next iteration.

What comfort there is to be derived from the comparatively more interesting alternative offered by this cyclical version of the Universe's fate was tempered slightly by the obvious fact that life as we know it could never survive the process of regeneration. For some, however, the possibility that the Universe was caught somewhere in the midst of a continuous series of bangs and collapses offered renewed hope of avoiding the need for it to have a unique beginning.

The factor that determines what fate the Universe will eventually face is (give or take a few technical clarifications) essentially just the balance between the total gravitational pull inwards of the Universe's contents and the outward expansion of space. This balance is often expressed in terms of the 'curvature' of space and time. An open Universe is one which expands forever, whereas a closed Universe will eventually collapse back in on itself. On the borderline between the two, a flat Universe is one in which the expansion rate slows and slows but never quite stops. In the virtual world of mathematics, a flat Universe stops expanding after an infinite amount of time has passed; in reality of course this end state is never actually reached.

As the final years of the millennium played out, there was little conclusive evidence either way to say whether the Universe would expand forever or collapse back on itself, but an exciting new piece of the puzzle was about to be put in place.

By the mid-1990s, our ability to repeat the kind of distance versus velocity measurements Hubble had made in 1929 had come a long way. Not only did researchers have much more precise knowledge of a larger number of so-called 'standard candles' (objects of known intrinsic brightness that could be used to determine the distances to galaxies) but advances in telescope technology also allowed astronomers to observe a far larger number of objects in the same study.

In particular, large scale surveys of a particular category of exploding stars, known as 'type Ia supernovae', in distant galaxies allowed scientists to study the expansion of the Universe in more detail and over a longer period of its history. The results of these surveys appeared to be showing something intriguing: the best fit to the expansion profile through time could only be obtained if an additional parameter was added to the mix, which became known as the 'cosmological constant'.

The cosmological constant is famous as having first been introduced as an idea by Einstein, who then abandoned the concept and proclaimed it to be his greatest blunder, only for it to be resurrected decades later as a result of the supernova studies and become a core feature of our model of the Universe.

In the early days of relativity, Einstein knew that the laws he had derived implied that space and time were dynamic entities. At the time, however, space was believed to be static, so Einstein fixed the equations by adding an additional element that counteracted the dynamism and prevented space from changing. This new element was the cosmological constant.

Given the elegance of relativity, the cosmological constant was a somewhat clunky way of mending what Einstein perceived to be a problem, demonstrating that even the greatest of theorists remain subject to their own prejudices and sense of aesthetic. Once the expansion of space was discovered, Einstein quickly disowned the arbitrary addition he had made to his equations.

Visionary genius though Einstein was, the fact that the cosmological constant came back does not mean he had gained insight to something that no-one else at the time could foresee. For the cosmological constant uncovered by supernova studies did not act to balance the dynamic nature of space; it only made things worse. The supernova research suggested that this addition was actually driving the Universe's expansion to accelerate. The source of the acceleration was unclear but was referred to using the coverall description of 'dark energy'.

Dark energy added even further mystique to a story of the Universe's history and evolution that was already extraordinary. While it provided a consistent explanation for the general characteristics of the cosmos, however, the tale also begged questions of its own. What was the particle that caused inflation, and how did it stop? How could theoretical models of the early Universe be tested? What was the source of dark energy? The answers to these questions were not immediately obvious, yet the more studies were made, the more the basic features of its narrative seemed to be reinforced.

The field's crowning glory came during the first decade of the new millennium when the follow up missions to COBE, called WMAP and Planck, determined to a much higher degree of accuracy the statistical distribution of variations in the CMB. This established gold-standard estimates of a set of parameters that could be used to model the expansion history of the Universe. When compared with the independent results of other cosmology projects, like the supernova surveys, the measured values of these parameters all appeared to be approaching agreement.

This was the golden age of observational cosmology. The near magic of being able to study the Universe at different epochs, and the enormous technological and methodological advances that enabled scientists to see further and deeper than ever before, meant that virtually the entire history of the Universe was laid out for us to examine, chapter by chapter.

So much progress had been made, but there was a lot of work still to be done and the potential for further major discoveries remained high. The excitement was intense, as scientists continued to nail down the essential values that described what came to be known as the 'standard model of cosmology', or 'concordance model'.

The problem with so much excitement is that one gets greedy. After a decade of quick-fire excitement as new updates of data were released the new discoveries began to slow. The scientific literature was awash with an abundance of theoretical models that sought to probe even deeper into the early Universe - in some cases even before the era of inflation - but with little hope of testing many of them, the overall picture ended up more confusing and harder to access.

Within theoretical cosmology there even emerged areas of study - in particular those related to the field of string theory, which seeks to provide a unified explanation for the behaviour of the quantum world as well as for gravity - that were all but inaccessible to professional physicists who had not themselves specialised in the area for many years. This challenged the field's ability to retain its objectivity and ensure research could be subjected to rigorous review.

There were more mundane issues with cosmology's branding too. The consensus around the standard model that had at first seemed so invigorating now became familiar and perhaps even a little suspicious. The other rockstars of astronomy, in particular the search for extra-terrestrial life and space exploration, began to take more of the limelight.

After such a long bright period, cosmology had ended up in a bit of a rut. It was about time for a shake-up. The first serious rumours of trouble came in 2016. Something was wrong with the present-day value of the Hubble parameter.

The problem with the Hubble constant losing its status as a real constant over time is that it becomes a lot more complicated to talk about the things that are actually measured by astronomers.

The present-day value of the Hubble parameter is, as the name suggests, the value of the Hubble parameter as measured in the Universe today. This is a more concrete thing to talk about than referring generally to the Hubble parameter, so cosmologists tend to frame their equations and models in terms of the present-day value, then adjust its value depending on which epoch they are interested in to account for the history of the Universe's expansion.

The present-day value of the Hubble parameter is, however, a bit of a mouthful and becomes somewhat tedious to keep repeating, so scientists use a shorthand to refer to it instead. Ordinarily this shorthand is either $H(0)$ or H_0, where the presence of the '0' distinguishes that we are talking about the present-day value. As a snapshot of the value of the Hubble parameter, H_0 really is a constant inasmuch as the present-day value should remain the same wherever we are in the (present-day) Universe.

Regardless of terminology, however, H_0 has become one of the essential components of the cosmological model. The classical method for determining its value is based on Hubble's approach, by measuring the distance to nearby galaxies and comparing this to their recessional velocities.

There are a number of different ways to do this. The measurement of velocity is generally considered to be relatively straightforward. It depends on breaking down the light from the galaxy into its constituent colours, or 'spectrum', and locating dark features called absorption lines, or narrow peaks of enhanced brightness called emission lines.

As the name suggests, absorption lines in a galaxy's spectrum are the result of light generated from its constituent stars being absorbed as it travels through gas in its interior. Similarly, emission lines are produced as a result of gas in a target galaxy first becoming excited in some way, then emitting light as the atoms within it get rid of their extra energy and settle back to their original states. The exact position in the spectrum where an absorption or emission line is found depends on the composition of the type of gas doing the absorbing or emitting. Based on studying the lines produced in controlled experiments on Earth, early researchers were able to recognise the patterns produced by different elements to work out what gases were present in the spectra of galaxies that were being observed.

Nowadays, studying spectral lines is routine. In galaxies that are moving away from us, however, the position of the lines get shifted. They are still identifiable, but fall in a slightly different place, toward the lower energy (redder) end of the spectrum. By working out the amount by which the lines have shifted, the 'redshift', scientists can determine the speed with which the galaxy is moving away from us.

The measurement of distance is somewhat more involved. A common method is to find standard candles in galaxies of known redshift and compare their apparent brightness with their inferred intrinsic brightness to calculate their distance. This is the method used in supernova searches as well as for Hubble's original studies in which he used Cepheid variable stars.

Along similar lines, the intrinsic luminosity of certain types of galaxies have also been found to correlate with the rotation characteristics or velocity distribution of the stars contained within them. For the classic spiral galaxies, this relationship is referred to as the Tully-Fisher relation, whereas for elliptical galaxies it is called the Faber-Jackson relation.

These approaches build knowledge of H_0 from direct measurement, by calculating its value during recent epochs and effectively extrapolating the observed relationship to the present day. They are, however, limited by our understanding of the method of determining distances that underpin them.

Entire professional careers have been dedicated to understanding exactly how 'standard' different kinds of standard candles really are and calibrating their intrinsic brightness in ever more precise ways. The task is a complicated one. In the case of type Ia supernovae, for example, the standard explanation for what triggers the detonation is based on a model of mass being transferred onto a so-called 'white dwarf' star from a companion in a binary system. When the mass of the white dwarf reaches a critical level, this triggers a runaway fusion process that leads to an explosion. The fact that there is a precise limit to the white dwarf's mass at which it detonates is the reason that this category of stellar explosion has a standard brightness.

One can imagine, however, that there could be all kinds of other complicating factors that could affect exactly how bright the explosion appears. The exact makeup of the star might influence the way in which energy is released as it detonates, for example, or the configuration of dust surrounding the binary system could block some of the light, making it appear dimmer. These factors are not easy to control for yet could change very slightly the brightness of the star when it explodes or its apparent brightness from our perspective on Earth. Similar kinds of problems affect other standard candles and the Tully-Fisher and Faber-Jackson relations. As a result, the observational error in the determination of H_0 for these kinds of studies has tended to be a significant limiting factor.

An alternative to calculating the distance to an object based on knowledge of its intrinsic brightness is determining its physical size and comparing that to its apparent (or angular) size in the sky. Similar to standard candles, objects that have a standard size are referred to as 'standard rulers'.

The angular measurement method was actually the approach used in studies of the CMB to calculate the distance to the 'surface of last scattering'. This surface defines all the locations in the Universe from which the CMB light is only just reaching us, which you can imagine as an enormous shell, tens of billions of light-years in diameter, centred on the Earth.

When the anisotropies in the CMB were measured by WMAP and Planck the resulting map of the sky contained features with a range of angular sizes in the sky, from tiny through to large. The researchers measured the amount of energy being received each second from variations at different scales and plotted this data to form a so-called 'power spectrum' for the radiation. The result - now famous among the astronomy community - clearly showed that the majority of the energy was coming from variations of a particular angular scale.

The clever thing was that this angular scale could be related to physical properties of the early Universe, which were calculated based on the fundamental theory for how the anisotropies were formed. These calculations allowed a physical size to be associated with the angular scale at which the energy being received peaked, which in turn allowed the distance to the surface of last scattering to be determined.

Knowing the distance and the redshift of light from the CMB then allowed the value of the Hubble parameter to be calculated at the time its light was released. In contrast to the direct measurement approach however, this only provided a single calculation of the Hubble parameter. The present-day value H_0 had to be extrapolated based on the other parameters in the cosmological model.

In effect the researchers were winding the clock forward on their models of the Universe from the time at which the CMB was generated to the present day, accounting for all the known physics that determines its evolution as they did so.

The problem that began to emerge in 2016 was that the consensus between the extrapolated measurement of H_0 based on anisotropies in the CMB and those obtained through direct measurement approaches - the same consensus that provided the original basis for the concordance model - had started to crack.

Truth be told there had always been some level of disagreement between the values calculated by both approaches, but the values had previously seemed close enough that the differences could be explained as being due to errors in the measurements.

Over the years however, understanding of the known errors in calculating H_0 based on supernova studies had continuously improved, and the value still did not match the one from CMB studies. On the contrary, the discrepancy between the two measurements was only getting worse. By 2019, it could no longer be ignored. There seemed to be only two alternatives. Either there was an unknown and significant source of error in supernova studies, or our understanding of the physics underlying the Universe's evolution - which allowed researchers to fast forward from the value of the Hubble parameter at the surface of last scattering to today - was wrong in some way.

At last, it seemed as though some excitement was returning to cosmology. Problems in science are good. They point to the fault lines in our knowledge of physics that we can investigate and seek to unravel. A betting person would have assumed that the error had to lie on the side of the complex supernova studies. At the end of 2019 and into 2020 the needle of responsibility for the differences in H_0 began to swing in one direction, but it was not what the scientific establishment had expected.

First, there were the direct measurements of H_0 based on gravitational lensing. These relied on the principle that the path of light travelling near a massive object will be diverted by its gravitational field, similar to the way in which light passing through an ordinary glass lens is refracted and ends up coming together at a single point of focus.

A common source of gravitational lensing in astrophysics is when light from an object in the distant Universe passes along the same line of sight as a nearby massive cluster of galaxies. Instead of the background galaxy just being blocked entirely, some of the light travelling past the cluster in a direction slightly away from the Earth gets diverted toward us. There are many beautiful pictures of rich galaxy clusters surrounded by the distorted images of more distant background galaxies that form ghostly shapes or elegant arcs of light at various locations in the field.

Importantly, gravitational lensing tends to produce multiple images of the same background object. The measurement of H_0 was made by observing gravitationally lensed images of distant quasars - point-like sources that are among the most luminous objects in the Universe - in six different galaxy clusters. In each of the clusters the background quasars varied in brightness, and this variation was seen to be repeated in each of the lensed images. Because the light that formed each image followed a slightly different path through space, however, the variation in their light also occurred at different times.

Measuring the time separation between the variation in each image allowed the team to work out what the alignment must be between the quasar, galaxy cluster and the Earth, and place limits on the ratio of the distance from the background quasar to the galaxy cluster, and from the cluster to Earth. While not measuring the distance to the quasar directly, this ratio could also be used to constrain the value of H_0. The conclusions were clear, and they agreed with the results from the supernova surveys rather than those of the CMB studies.

Next, in February 2020, an international collaboration using some of the world's most advanced radio telescopes and telescope arrays peered deep into the hearts of four galaxies to study edge-on discs of material swirling around their supermassive central black holes.

Located within these ferocious gravitational whirlpools were sources called masers - the microwave equivalent of lasers - generated by water molecules. The masers' speed of rotation helped build a picture of the kinematic properties of each disc, which could be used to determine its physical size. The distances to the host galaxies could then be calculated by comparing the physical sizes of the discs to their apparent angular sizes in the sky. Once again, the resulting value of H_0 was consistent with the supernova studies.

Then, in July 2020, the distances to almost one hundred galaxies were measured using an updated version of the Tully-Fisher relationship, which had been refined and re-calibrated. When compared to the galaxies' redshifts, the calculation of H_0 was even further from the CMB value. The measurement was however still consistent with results derived using the other direct measurement approaches.

By the time the European Space Agency's Gaia mission released new results at the end of the year that reduced the errors in supernova studies even further, there could no longer be any doubt that the crisis in cosmology was real.

Crucially, each of the direct measurement studies used a different approach to measure distances yet drew similar conclusions. This made it very unlikely that an unaccounted-for error in the observations was to blame for the lack of agreement with CMB studies. The error would have to be common to each of the different techniques in order for it to explain their general agreement.

In an ironic twist, the concordance model - which was built on the apparent consistency between different measurements of its essential parameters - was being undermined by a new consensus.

The matter is still far from being decided, but there is a growing body of scientists who now believe that the source of the discrepancy in the measurement of H_0 has to be a result of something missing in the cosmological model. This suggests there may be new physics that needs to be accounted for in the description of the Universe's expansion.

Of course, it has not taken long for speculation to begin on what this new physics might be. One suggestion is that the simple characterisation typically assumed to describe the influence of dark energy on space may be wrong, or that there is an unknown interaction between dark energy and dark matter that affects the Universe's expansion. Another option is that the early behaviour of the Universe may have been affected by the presence of exotic particles, or as a result of changes in the properties of one of nature's most diminutive particles, the tiny neutrinos. No doubt there will be many more proposals made before we have enough data to be able to distinguish between them.

Meanwhile the intrigue is growing and there are whispers of revolution that may prove infectious. While the essential regime of the Big Bang model is by no means in question, there are persistent rumours of scientists losing faith with the paradigm of inflation, accusing it of having lost its predictive power and, as a result, its scientific credentials.

For the time being, these arguments are mostly confined to a more philosophical level of debate. There has however been a growing sense of unease around the lack of direct observational evidence that supports the theory, despite it being over fifty years since it was first proposed. One of the few predictions made by inflation is that the early Universe would contain ripples of space and time known as 'gravitational waves'. These ripples would be detectable by the effect they have on the orientation of light in the cosmic microwave background. So far however this so-called 'polarization signal' has eluded the studies searching for it.

The rapidly narrowing landing strip for detecting primordial gravitational waves predicted by inflation also acts as a reminder of those other nagging questions for which cosmology has yet to find a secure foothold. Despite decades of research, for example, we are almost no closer to identifying what most of our Universe is made of. Astronomers have long believed that the majority of the material making up galaxies and galaxy clusters comes in an unseen form known as 'dark matter', but we are still do not know what this enigmatic substance is. The nature of dark energy - which constitutes an even greater fraction of the overall mass of the Universe - is even more mysterious.

As a result, the entire edifice of our cosmological model walks a continuous tightrope. It is a story of singularly impressive scope and detail, built and held together by the enormous effort and dedication of the scientific community, yet one that is always seemingly only mere moments away from falling apart.

Of course, to be in this position is also exhilarating. It is one of the beauties of studying the great patterns of the cosmos that further fundamental discoveries are always only one unusual result away. The discrepancy in H_0 may, with time, prove the key to unlocking and improving our understanding of cosmology. If nothing else, it has reinjected much-needed excitement into the field. For the time being, at least, cosmology has got its mojo back again.

Earth's
Unlikely Twin

I n occasional moments of whimsical fancy I sometimes find myself imagining the solar system to be a kind of mythical celestial family born, as if by magic, from the fertile dirt of space. At its centre sits the Sun, resplendent in her swollen maternity, providing comfort and chastisement to the eight major planets that spin and dance around her, as well as to the vast panoply of minor planets, asteroids and comets that make up the rest of her many other children.

As with any family, there are many similarities and differences between the solar system's siblings. Perhaps the most obvious subdivision exists between the four planets that orbit closest to the Sun and those that make their way silently through space further out. The dividing line between the inner and outer planets of the solar system is marked by the asteroid belt; a broad band of small, irregularly shaped rocky objects that orbit roughly two to three times further from the Sun than the Earth does.

Located beyond the border of the asteroids are the planets Jupiter, Saturn, Uranus and Neptune. These are the heavyweights of the Sun's offspring - the gas giants, made up primarily of hydrogen and helium. The largest among them is Jupiter, with a diameter just shy of 140,000 kilometres and a mass of around two million, billion, billion tonnes. The planet is particularly distinctive for its multicoloured filigreed bands of clouds and of course for its red spot - a storm the size of the Earth that has been raging in Jupiter's upper atmosphere for centuries - which peers out into the rest of space like a baleful eye.

By comparison, Saturn is only slightly smaller than Jupiter, but is famous for its majestic ring system, the main sections of which extend up to 130,000 kilometres away from the planet. Its disc is a pale yellowish colour covered with a more subtle pattern of bands than those that decorate Jupiter.

Located further out from the Sun, Uranus and Neptune are both less than half the size of Jupiter and Saturn and by comparison are relatively featureless. At first glance both look like watery worlds - Uranus a near featureless orb of serene green-blue, and Neptune a much deeper blue, flecked with occasional white. In reality however the bluer tones of both planets are caused by absorption of red light, in particular by methane in their atmospheres.

Each of the outer planets have their own extensive families of satellites, some of which are believed to have formed out of the same material as their guardians, while others were originally strays that ventured too close to the giant planets and ended up ensnared by their gravitational embrace.

There could hardly be a greater contrast between the gas giants beyond the asteroid belt and the four inner planets. Mercury, Venus, Earth and Mars are all much smaller and less massive, and are distinctive for having rocky surfaces made up of heavy elements and relatively thin atmospheres.

Of these, Mercury is the innermost planet. A small, cratered world, its surface at the equator scorches during the daytime as a result of its proximity to the Sun, while temperatures during its night plummet to well below -100 degrees Celsius. Mercury is often unfairly written off as the runt of the family, characterised as a lifeless, relatively uninteresting world.

The Earth, on the other hand, is arguably the most interesting of the inner planets. This is not just anthropomorphic bias. Our world is the largest, densest and heaviest of the four inner planets and is the only one with a substantial-sized natural satellite - the Moon. It is also the only planet which has a substantial proportion of its surface covered with liquid water and, yes, it is the one place in our entire Universe known to support biological life. This simple fact should be sufficient enough cause for interest, regardless of the fact that humans count as one of the many species that also call Earth home.

In any case the contrast between the Earth and Mercury and - albeit for different reasons - between the Earth and the gas giants is pretty stark. If one was to ask therefore, which of the other planets in our solar system most closely resembles the Earth, only Venus and Mars really remain in contention. For many years, however, when people have spoken of the most Earth-like planet in the solar system, it is the red planet Mars that they are talking about.

It is not hard to understand why. Not long after the first telescope was directed at the night sky, astronomers noted the existence of dark features on Mars' surface; white regions that looked something like ice caps, and even yellowish clouds reminiscent of dust storms. In contrast to the other planets, Mars looked as though it had features that could be reminiscent of the landscapes on Earth. As telescopes improved and more detail was revealed of Mars' surface, some of the dark regions even appeared to be arranged in long, straight lines that looked artificial, similar to man-made canals.

There are few things more enticing to the human imagination than suggestion. The apparent similarities between Mars and Earth proved potent fuel for the idea that the planet may also harbour life. That fuel was stoked by the publication of H. G. Wells' classic 'War of the Worlds' in 1897, which even envisioned Martian lifeforms that were more advanced than us.

While the canals on Mars were eventually found to be optical illusions, there is now overwhelming evidence that between about 4 billion and 3.5 billion years ago there was a significant amount of water present on Mars and that its terrain was sculpted by the eroding action of flowing water or melting snow. There is even direct evidence that rain fell on its surface, in some cases leading to flooding, supporting the idea that Mars hosted an active weather system as well.

It is difficult to estimate exactly how much liquid water may have covered the surface of the planet, but current studies have identified several hundred dried up lake beds of comparable size to ones on Earth. The total amount of ice and snow detected on Mars so far would be enough to cover the whole planet with a shallow sea. Some scientists have even hypothesized that large areas of lower terrain in Mars' Northern hemisphere may have been the site of an ocean.

The reason everyone gets excited about liquid water, of course, is because of its importance for life as we know it. This is a result of water's molecular structure, expressed by its well-known chemical formula, H_2O. This molecular shorthand tells us that water molecules are composed of two Hydrogen atoms (the 'H_2' bit) and an Oxygen atom (the 'O') bound together.

What is not immediately obvious from the formula is that when water forms, the individual atoms bind in such a way that the Hydrogen atoms end up mostly on one side of the molecule with the Oxygen atom on the other side. This arrangement results in a slightly uneven distribution of charge across it, meaning that water can develop weak bonds with a wide range of other molecules. In chemistry-speak, it is an excellent solvent within which other molecules easily dissolve. This makes liquid water an ideal medium for transporting the essential nutrients and minerals required for biological functions across cell membranes to where they are needed.

A less obvious, but no less crucial, property of water is that it transitions between solid, liquid and gas states over arrange of temperatures and pressures that are also favourable for life. This is important because water does not just find its way to where it is needed based on its own initiative. Instead, there need to be physical processes in place that ensure water gets moved around, and this is where transitions between different states prove helpful.

When I was young, I remember learning about the water cycle in school. I learned that a proportion of the liquid water from the land and the oceans is heated by the Sun and evaporates, changing into a gas, and rises into the atmosphere. There it cools once more and condenses on tiny grains of dust or ash to form small droplets that are light enough to remain suspended in air. Large collections of these droplets end up forming clouds that move around the globe, hurried along by winds.

These nomadic clouds continue their travels until the water within them condenses into even larger drops, perhaps as a result of passing over hills and mountains where the rising terrain forces clouds higher again into regions where the air is still colder. Eventually the drops become so heavy that they can no longer be supported in the air and begin falling out of the sky as rain or hail or snow.

Where it falls on land, the liquid water runs over and through the soil until it either freezes, evaporates back into the air, or finds its way to underground reservoirs, rivers or oceans. Water that freezes as snow or ice may eventually warm back up again to become liquid, or else is transported elsewhere in the midst of the glaciers that grind their way through the rocks of mountains, slowly carving out the land to provide new opportunities for life in the midst of Earth's ongoing cycle of reinvention and renewal.

What I did not understand when I first learned about the water cycle was the fundamental way in which it was essential to the existence of life on Earth. On its way back to the oceans or the air, for example, the water that runs through the soil can be absorbed through the roots of plants or drunk by animals where it is used in a huge multitude of biological processes, as well as carrying essential minerals and nutrients into the waterways, supporting aquatic life.

While there are other molecules that act as good solvents, none are as general-purpose as water. Alternative solvents also tend to be less abundant and do not have the same versatility in terms of the temperature at which they change phase. As a result, the existence of liquid water is considered to be virtually a prerequisite in the identification of places that could support life beyond the Earth.

Of course, simply finding liquid water does not mean that life will also exist. Happily, however, researchers are not immune to their own dreams and fantasies. The fact that life on Earth is so abundant as a result of its ample supply of water makes it tempting to believe that wherever liquid water exists, the odds that it will be accompanied by new forms of life are pretty high. As a result, there remains a high level of expectation that when Mars was warmer and wetter it may also have hosted life.

Conditions on Mars now are, of course, quite different. The planet's meagre atmosphere is insufficient to protect water atoms being broken apart by high energy ultraviolet radiation from the Sun, and its weak gravitational field is not strong enough to prevent vapour being lost directly into space. What remains of the water that previously carved out channels in the Martian landscape is believed to be either frozen in its soil, locked into its ice caps, or is otherwise stored below its surface.

While Mars no longer has enough liquid water to support a diversity of life however, it remains possible that highly adapted microbial life could still survive there.

For over twenty years, Mars has offered small but significant scraps of evidence that have helped sustain this dream. Scientists have long been aware, for example, of plumes of methane in Mars' atmosphere. Methane is a molecule that is easily broken down by solar radiation, so the existence of these plumes suggested that the gas was being actively produced somewhere on or within the planet. Methane has subsequently been detected by both the European Space Agency's Mars Express satellite and NASA's Curiosity rover. The Curiosity rover even discovered seasonal variations in levels of the gas.

One possible explanation is that the observations are due to structures called methane clathrates in the subsurface, within which methane molecules are trapped into ice crystals but leak out as the weather changes and the crystals melt. The more attractive alternative, however, is that the methane is a by-product of biological processes in microbial colonies in the Martian subsurface. As yet, however, scientists are unable to tell definitively which of these answers is correct.

Even more exciting has been the discovery that liquid water still exists in small pockets on Mars today. In 2011, NASA released results from their Mars Reconnaissance Orbiter based on studies of narrow dark lines on the slopes of crater walls and Martian highlands. At first glance, these features seemed reminiscent of small rockfalls, but the researchers found that they were not permanent. Instead, they appeared during the relative warmth of Mars' Summer period and disappeared during its Winter. This seasonal variation suggested that these so-called 'recurring slope lineae' could be caused by melting water lubricating the Martian soil and leading to flows of material downhill.

This conclusion appeared to have been confirmed in 2015 with the detection of hydrated salts on the same slopes where recurring slope lineae were observed, which appeared and disappeared with the same frequency as the lines themselves. As the name suggests, hydrated salts are composed of salt molecules bound loosely to a number of water molecules and crystallise out from salty water as it evaporates.

The interpretation seemed clear - as the warm Martian summer days begin, recurring slope lineae were created by salty liquid flows trickling down the slopes. Eventually however the water in the flows evaporates, ending its movement and leaving behind a calling card in the form of hydrated salts.

The most sensational news came in 2018 however, when researchers from the Mars Express announced that they had discovered a lake of liquid water several tens of kilometres in length, located about a kilometre and a half under Mars' South Polar ice cap. The detection was confirmed in further studies announced in September 2020, which also found three smaller lakes located nearby.

Several hundred subglacial lakes are known to exist on Earth, most of them under the Antarctic ice cap. The largest of these is Lake Vostok, a huge body of freshwater almost 250 kilometres long trapped several kilometres below the surface. In fact, the majority of subglacial lakes on Earth are believed to be freshwater and support an incredible assortment of life that has evolved to survive the extreme conditions within them.

Unfortunately, in the case of the Martian lakes the subsurface conditions would not normally be expected to support water in a liquid state. The fact that liquid water is observed, therefore, has led scientists to conclude that the water within them must be extremely salty. Increasing the saltiness, or 'salinity', of water lowers its freezing point thus allows it to remain in liquid form at temperatures well below zero degrees Celsius. At first glance, this might seem to be a good thing for life. The downside, however, is that high levels of salt in water tend to reduce the availability of the water for use in biological processes. As a result, saline environments are generally extremely bad for all but the most adapted forms of life.

While Mars continues to tantalise, therefore, none of the discoveries made so far have provided firm evidence that life ever existed there or still exists today. All the elements of the story seem to be in place however and the dreamers still have reason to hope that it is only a matter of time before we find biology's smoking gun in the soil or rocks of the red planet. As a result, Mars remains one of the most obvious places to look for life existing beyond our home planet. The same cannot be said, however, of Earth's other nearest neighbour in the solar system, the planet Venus.

Among the regular occupants of the night sky, Venus is the brightest object after the Moon. Its position within the orbit of the Earth means that it tends to remain fairly close to the Sun in the sky. As a result, Venus is often a prominent object in the early hours of the night or in the lead-up to daybreak. Its purity and brilliant shine, so often found cradled within the soft folds and sublime lighting of sunset or sunrise, makes Venus the inspiration of many a romantic folly.

It is no wonder therefore that ancient cultures across the world named the planet after their deities of love and beauty, or have associated it with clarity, greatness or reproduction. Little did they know that the real nature of Venus was about as far from affectionate or amorous as it is possible to be.

Despite being a familiar object for millennia, very little was known about Venus until late in the 1960s. Early observations using telescopes and during transits - periods when the planet passed directly in front of the blazing disc of the Sun - revealed only that it had a thick atmosphere through which nothing of the surface could be seen.

Fuelled perhaps by speculation about life on Mars, Venus' atmospheric veil inspired theories that, hidden beneath the clouds, there was a climate not that much different from Earth. Indeed, before Mars stole the crown, Venus was a much more likely twin to Earth. The two planets have about the same size and mass and, as Venus is the closest of the planets to Earth, they also likely formed in roughly the same environment in the solar system and out of similar material.

With its cloud cover providing an additional level of insulation, it was conceivable that Venus' surface temperature could also be very similar to Earth's. It was not until the Soviet Union fired the first probes in the direction of the planet, only to find them brutally crushed as they plummeted into its thick atmosphere, that the reality began to be uncovered.

In 1967, the Soviet probe Venera 4 survived long enough in the Venusian atmosphere to perform measurements of its temperature and chemical composition. Results from the tiny, doomed probe showed that the temperature near the planet's surface was around 500 degrees Celsius - hot enough to melt lead - while the atmosphere was composed almost entirely of carbon dioxide, with little or no trace of water.

Today, any hope of life existing on the surface of Venus has long since been quashed. The smothering blanket of its heavy atmosphere keeps the temperature unbearably high; the pressure is around 90 times greater than at sea level on Earth, and what appear like tranquil clouds in high resolution images of the planet are composed of droplets of sulphuric acid rather than water. If you or I were transported to Venus, the only debatable mercy is that we would probably be crushed faster than it would take us to die by being roasted alive or asphyxiated.

Little wonder therefore that it was not long after the horrific conditions at its surface had been confirmed that the broad beam of space exploration's searchlight swung away from Venus in the direction of other parts of the solar system. The most enduring legacy of the early exploration of Venus was arguably the warning it served humankind about the potentially devastating consequences of forcing a climate out of an otherwise benign equilibrium.

The question was why two planets such as Earth and Venus, with such similar properties and located in such similar places, could have ended up so radically different. In one sense the answer was obvious - the carbon dioxide that makes up the vast majority of Venus' thick atmosphere is a greenhouse gas, which serves to retain heat from the Sun and prevent it escaping into space.

On Earth, relatively low levels of greenhouse gases in the atmosphere maintain - for the time being - a comparatively balmy temperature on the surface. On Venus, they shroud the planet completely, allowing the temperature to rise to unsufferable levels.

What was not so obvious was why this huge difference in atmospheric carbon dioxide concentrations should exist in the first place. Clearly, if Venus began life as a planet similar to Earth, it must have suffered some kind of runaway process that allowed carbon dioxide to build up. For a long time, the source of that greenhouse effect seemed to have something to do, yet again, with our familiar friend, the humble water molecule.

The temperature of Earth's atmosphere is maintained over time by a complicated but elegant process known as the Carbon cycle that shifts carbon around between the atmosphere and the land and oceans, eventually burying it under the Earth's surface before it is released again as a result of volcanic activity. Under normal circumstances, this cycle manages a delicate balance in the amount of carbon dioxide in the atmosphere and, therefore, the degree of insulation our atmosphere provides.

One of the steps in the carbon cycle on Earth is a reaction that takes place between carbon dioxide gas and rainwater in the atmosphere, which produces a weak form of acid rain. When this rain falls, carbon is removed from the atmosphere. Over time the acid erodes rocks on the land, releasing key ingredients for the next stage of the cycle which are then swept into the oceans.

The original story of how Venus became so hellish therefore, was simply that water vapour was lost from its atmosphere as a result of its being located closer to the Sun and receiving larger amounts of high energy ultraviolet light as a result. This process works in two stages. First, the radiation breaks down water molecules into their constituent hydrogen and oxygen atoms. Then the hydrogen - which is the lightest of the atoms - ends up escaping the atmosphere, thereby preventing the possibility of the water molecule reforming. As a result, it was believed that the amount of water vapour in Venus' atmosphere would have reduced relatively quickly.

This gradual breakdown of the planet's atmospheric water would have slowed the rate at which carbon dioxide was removed through acid rain, allowing it to build-up instead. The resulting runaway greenhouse effect would have been boosted by the fact that water is itself a greenhouse gas, so as the early oceans on Venus evaporated - again a result of its closer location to the Sun - the resulting vapour would have also generated a warming effect during the short interval of time that it survived before being chipped apart by solar rays. This would have exacerbated what was already a bad situation. Once the carbon dioxide took over entirely, the planet was doomed to end up a furnace.

Up until the mid 2010s, the prevailing assumption was that this process would have happened in a relatively matter-of-fact manner. If Venus did start off looking quite similar to Earth, therefore, there would probably be only a relatively limited window of opportunity for life to take hold.

Over the past five years or so, however, studies have begun to show that, with a little bit of luck, Venus may have been able to retain an Earth-like climate for much longer than expected. The models rely on specific assumptions about the rotation of Venus and the amount of water that originally covered its surface, but they demonstrate that it could have been possible for clouds to cool the planet by reflecting the additional solar heat back into space rather than trapping it close to the surface.

This would counteract the insulating effect of Venus' greenhouse gasses, slowing the evaporation of water and so reducing the rate at which it was lost as a result of being broken down by the Sun's ultraviolet radiation. There are even models in which Venus could have remained habitable right up to the present day. Clearly, however, this is not what happened. If the models are right, therefore, this suggests that the change in Venus' environment from homely oasis to incinerating oven could have been the result of something more catastrophic than the gradual drying up of the atmosphere.

It is not yet clear what this catastrophe might have been. One possibility is that Venus could have collided with an object with enough force that the shielding atmosphere of the planet was completely stripped away during the impact. The energy released could also have melted or cracked through the crust of the planet, releasing vast quantities of carbon dioxide stored below its surface that went on to form the planet's new atmosphere as it cooled once more.

Another idea is that Venus was just geologically unlucky. Volcanic activity on Earth constantly releases carbon dioxide into the air, but the volume of emissions is generally quite moderate. By comparison, humans are responsible for around 40-100 times more emission than all the volcanoes in the world combined. We also know, however, of a number of occasions during the past 500 million years where volcanic emissions have apparently increased significantly, leading to rapid global warming. Maybe Venus was hit with a string of such events that swung the balance of carbon dioxide in its atmosphere too far away from equilibrium, such that it was unable to recover.

Scientists are not without options for what might have caused Venus' environmental catastrophe. There is, however, another possibility - one which is perhaps even more dramatic and appropriate considering the tortured world Venus has become. Clues as to what this may have entailed can be found in the geology of Venus' landscape.

It is over 30 years since NASA launched its last probe dedicated to studying our nearest planetary neighbour. When the Magellan spacecraft was blasted into space in May 1989, it was with the aim of using radar to peer through Venus' cloud layer and chart the terrain below. This would be the most comprehensive view yet of the hidden world and scientists were anxious to discover what it would find. The mission did not disappoint. Over the course of two years, Magellan methodically mapped almost the entire globe, revealing the Venusian surface in unprecedented detail.

It quickly became clear though that there was something unusual about the landscape Magellan uncovered. Venus had lots of volcanoes, but most seemed to be inactive and there was no evidence of Venus having tectonic plates. The volcanoes were also quite different to those on our planet. Rather than the mighty jagged peaks we would normally imagine, crowned in angry clouds of vapour and ash, the volcanoes on Venus were shaped like vast domes, whose flanks rose only very gradually over hundreds of kilometres. If you were to hike up one of these so-called 'shield volcanoes', you would barely notice you were climbing any higher in altitude until you reached the summit.

As a result, Venus' landscape was generally relatively flat. It was also unusually smooth. Most rocky bodies in the solar system bear the scars of numerous impacts with other objects. Yet Venus had relatively few craters in comparison with the other inner planets and the Moon. The reason for this lack of cratering is not because Venus managed to dodge the bullets that struck other planets. Instead, it is believed that the entire surface of the planet is relatively young, most likely of the order of 600 to 800 million years old. It is as if Venus somehow recycled its crust and developed a fresh new skin in its place.

The exact mechanism that could have caused this resurfacing remains a matter of debate. One suggestion is that it was caused by the cessation of plate tectonics. In this scenario the planet may have been able to support an Earth-like habitat for a long time, but still lost enough water that the planet's crust began to dry out.

Quite apart from all the other things it is useful for, water is also believed to be essential for sustaining plate tectonics because it increases the viscosity of a planet's crust, weakening it and preventing it becoming strong enough to resist the movement induced by molten currents from within a planet's interior. As Venus' crust became parched, therefore, this may have slowly switched off the action of plate tectonics.

This is a problem not only because plate tectonics play their own essential role in the carbon cycle (storing carbon locked into rocks back underground through subduction zones, as well as enabling the release of carbon back into the atmosphere through volcanic activity), but they also provide a route for heat generated by radioactive decay deep within the planet's core to escape.

If Venus lost its tectonic activity therefore, its interior could slowly heat up, eventually becoming so hot that subsurface magma would actually begin melting the crust from beneath. Eventually, the weakening crust would collapse and be replaced by a sea of lava. Over time, the magma from the interior would cool and harden once more, providing Venus with a new and youthful face.

Just as in the scenario of a giant cosmic collision, a catastrophic resurfacing event would also have released a huge amount of carbon dioxide that had been locked under Venus' surface into the atmosphere over a very short period of time, seeding the planet's subsequent runaway greenhouse effect.

Quite how things would have played out in the centuries leading up to such a resurfacing event is not clear. One can only hope that any life that existed on the planet had already died out long before it took place, rather than being trapped in what could only have been a terrifying inescapable nightmare.

There is still so much we do not understand about Venus that it remains difficult to distinguish which of the scenarios for generating its runaway greenhouse is the most likely to be true. The list of possibilities is also not exhaustive and future research is sure to offer other alternatives as well. Whatever the cause of Venus' young crust and its deadly atmosphere however, it seems impossible that anything could have survived such an environmental apocalypse.

It was therefore a huge surprise when, in September 2020, scientists analysing data from a telescope located on the summit of Hawaii's largest dormant volcano, Mauna Kea, announced that they may have found evidence of life existing on Venus.

The story goes that when Professor Jane Greaves from Cardiff University in South Wales originally decided to check for signs of life on Venus, the data she gathered sat on her computer for about a year and a half before she decided to analyse it in order to clear space for more observations. Clearly, not even she had much hope of making a breakthrough. So when the results of her analysis came back showing exactly what she had hoped to see, it must have come as something of a shock.

In a list of glamorous molecules, phosphine would never rank highly. It is a particularly toxic gas that has a nasty habit of causing pulmonary oedemas if ingested. When mixed with small levels of impurities, it is also said to smell badly and can spontaneously combust in air. It is also pretty reactive, meaning that it quickly disappears from an environment unless it is being replenished.

Despite its toxicity, therefore, phosphine detected on Earth has always been found to be associated with biological processes. The fact that it appears to be an unlikely by-product of some forms of biological activity means that phosphine has previously been suggested as a possible biosignature, whose detection in an alien world would raise a potential flag for the presence of life.

In fact, phosphine has already been detected in the solar system's gas giants but, in these cases, it is believed to be have been produced in their hot interiors then dredged upwards by the constant churning, or convection, of the planets' outer layers. Things are different for the rocky inner planets, where the hard crust of the planet prevents any phosphine generated within from being mixed into the atmosphere in the same way.

As a result, Greaves reasoned that Venus was an ideal place to search for phosphorus generated by organic processes in the solar system. If white light is shone through a gas, the atoms or molecules of that gas can absorb small amounts of that light at very specific wavelengths. This results in dark lines in the spectrum of the light known as 'absorption lines', which can be used to determine the composition of the gas.

Reflection of sunlight from the thick atmosphere of Venus would provide the perfect canvas upon which any absorption of light by phosphine would stand out. To Greaves' surprise when she processed the data from Mauna Kea, she found a dark line in exactly the place you would expect if phosphine was present.

Follow up observations quickly seemed to confirm the detection. What was more, they seemed to suggest that the phosphine was originating from a region around 50 kilometres above the surface of the planet. Scientists have long argued that, if there is anything resembling a sweet spot for life on Venus, this is precisely where it would be. At this altitude, the winds of the atmosphere are believed to be more moderate and the temperature and pressure are almost Earth-like. The major downside of this region is that it is also where a large amount of sulphuric acid droplets form, but these droplets harbour what little water there is left in the atmosphere, which might be accessible to airborne microbes (bringing new meaning to the saying that every cloud has a silver lining).

The chances have always seemed slim, but microbial life has been found to be extremely tenacious on Earth, existing in extreme environments that otherwise seem hopelessly hostile. As was true for Mars, the hope is that, while the majority of life that once populated Venus has long since gone extinct, highly adapted forms could still be clinging onto existence - lonely relics of an ancient and long-gone biological kingdom.

Besides, regardless of how unlikely it might seem, if the phosphine detection was real then it demanded an explanation. Try as they might however, the scientists could find no viable pathways to producing the phosphine signal as a result of non-biological processes on the planet.

It is credit to Greaves and her colleagues that, despite this, their reporting of the detection remained measured. The line that had been discovered was only one of a set of absorption features expected to exist if phosphine really were present. Further work would be needed to confirm that the signal was not a result of confusion with some other as yet unknown, chemical signature.

Not long after the detection was reported, independent reassessment of the data also suggested that the approach used to analyse the original observations might have been subtly flawed. If the new analysis was correct, the original results may not have been statistically significant after all, and the phosphine detection could have been nothing more than a ghost in the data.

There are yet more reasons to remain cautious. The biological routes for producing phosphine are still poorly understood, and the link between phosphine and life on Earth is based largely on association rather than knowledge of an underlying mechanism for generating the gas. Without this understanding, we have no way to be sure that those processes could be supported in any Venusian environments where phosphine is found.

As a result, the jury is still out on whether phosphine really does exist in the atmosphere of Venus. There is much that lies behind the pretty pictures and convincing graphics of popular science that is rarely seen or acknowledged. Identifying faint features in complex data is hard and often requires a number of critical assumptions that remain hidden in the details. If Greaves' initial analysis was indeed wrong, one of the issues will have been the way in which interfering noise was characterised and removed from the data to reveal the signature of phosphine absorption. This serves as a reminder of how difficult it is to detect unequivocal biosignatures in complex data.

Even if this ends up being the final verdict of the scientific community, the case is still not completely settled. It remains possible that phosphine exists, but in quantities that are too low to produce an absorption feature that is prominent enough to be confirmed reliably. If this is the reality, then the only way we are going to be able to tell is by sending new probes to Venus. Already there have been renewed calls for NASA to end its long drought of missions to the planet. If nothing else, there is still so much we do not understand about how Earth's unlikely twin ended up on its pathway to desolation. The mystery has remained outstanding for long enough and deserves a clearer answer.

Perhaps, however, the truth is that deep down we do not want to find the answer. Perhaps we do not want to face the possibility that, many hundreds of millions of years ago, Venus was a planet not unlike our own, supporting a rich diversity of biological life, that was swept away by environmental catastrophe.

If the stories about Mars and Venus are true, then the solar system started out with three habitable planets, which have gradually been whittled down to just one. On the one hand, that only emphasizes the uniqueness of our little blue world, but it also makes our home a lonely and vulnerable place to be. Perhaps we do not want the reminder of just how fragile our own habitat is, or how brutally it can wipe us away.

The Greatest Debate

I f there is one theme that most frequently attracts people to the subject of space, lighting the fuse of a lifetime's fascination, it is the humbling sense of perspective that it provides on our place in the Universe.

Often the feeling comes in response to the raw scale, power or beauty of the objects we observe. Whether it be the grand arch of the Milky Way straddling the night sky from a dark site; or the approaching rush of the Moon's shadow over the Earth's surface as it passes directly in front of the disc of the Sun, or the graceful intertwining of two merging galaxies acting out a collision of such inconceivable enormity that its progress can only be charted over many millions of years, there is nothing quite like the wonders of space to remind us of how tiny we are and how fleeting our time on the Earth is.

What is easily forgotten in these moments of transportation is that our sense of the immense scale of the cosmos, and the structures that exist within it, is actually still a relatively modern notion. April 26th, 2020 marked the hundredth anniversary of a meeting that took place in the grand surroundings of the

Smithsonian Museum of Natural History in Washington D.C. that was intended to tackle a fundamental question regarding the character of the Universe.

The meeting was billed as a debate between two leading astronomers of the time, Harlow Shapley and Heber Curtis, and was played out in front of an audience consisting of members of the public and academics said to have included Albert Einstein.

The topic for discussion was the nature of mysterious spiral shaped nebulae that had been observed by astronomers. On one side of the debate, Shapley was to argue that the Milky Way was the only major structure in the Universe, and that the curious spiral nebulae were minor objects located within it. On the other, Curtis would claim that the spiral nebulae were in fact objects separate to our own star system - so-called 'Island Universes', or 'galaxies' - located at enormous distances from Earth. The resolution of this question would change our perception of the cosmos forever.

Our understanding of the nature of the Universe and our place within it has changed radically over the past two thousand years. There is a tendency for this evolution to be represented as always progressive and always driven by enlightened Western thinking. The reality of course is more complex. Long before Galileo trained his telescope at the sky, for example, some early Greek philosophers and later Arabic scholars from the middle ages proposed that the irregular, dusty band of the Milky Way was composed of countless hosts of unresolved stars. The general view regarding the nature of the Universe among Western scholars up to the mid-sixteenth century was however, founded largely upon aesthetic concepts of the Universe's perfection.

Although religious ideals are often blamed wholesale for this subjective principle, its roots lie firmly in the philosophy of Plato and his student Aristotle in the fourth century BC, who developed the argument that the irregular and messy appearance of the physical world is an artifact of our flawed observation of it. According to this view, the underlying essence of reality was based on perfect mathematical forms, which reflected its true nature.

Plato and Aristotle were also proponents of the geocentric model of the heavens, in which the Earth lay at the centre of the entire Universe. It was already well-established that the Earth was shaped like a sphere rather than the flat plane of earlier traditions, and it was entirely in keeping with Plato's philosophy that the arrangement of the rest of the Universe should also be based on this most perfect of geometric forms. As a result, the first geocentric models were designed with the Sun, Moon, planets and stars moving on concentric spherical surfaces arranged Earth.

It was also well known, however, that the planets moved across the skies in ways that could not be explained by a simple circular motion. Sometimes they would shift position relative to the constant background of stars in a regular way; sometimes they would appear to stand still before doubling back on themselves, as if retracing their path across the heavens. To explain this motion, the geocentric model was extended so that the heavenly objects performed smaller circular loops called epicycles as they moved around the Earth on the surface of their spheres.

Over time however, the precision of observations of the locations of the stars and the relative motions of the planets were improved even further and by the time of Claudius Ptolemy in the second century AD, there was enough data to show that the epicycles were still not sufficient to explain the most detailed motions of the planets.

Ptolemy added a new set of circles to the model that gave much better predictions of planetary positions. This new regime also displaced the Earth ever so slightly from the exact centre of the planetary orbits, although we remained the centre of the Universe as a whole.

So comprehensive was his work that the Ptolemaic system became the basis for European understanding of the arrangement of the heavens for much of the next fourteen hundred years. Its reign was not entirely without challenge, in particular from Islamic scholars of the tenth to thirteenth centuries, some of whom even proposed that the planets might move on elliptical paths rather than following a complex network of perfect circles.

Nevertheless, Ptolemy's vision of the Universe endured. It was unique not only because the Earth was enthroned in the centre of all things, but because of the clockwork character of the spherical orbits and concentric domains of each of the celestial objects. Beyond the outermost sphere of the stars was the domain of heaven, making the Universe completely self-contained. There is something quite mechanical about this view of nature, as if the Universe were a finely crafted artifact suspended amidst the domain of heaven.

It was not until the early part of the seventeenth century that conviction in the Ptolemaic system began to be undermined. Copernicus had taken a tentative step toward change in the mid-sixteenth century with the publication of a so-called 'heliocentric' model for the Universe. Although this retained some of the features of Ptolemy's geocentric system - in particular that the movement of heavenly objects were based on circles - the major innovation often credited to Copernicus was that the Sun occupied the centre of the cosmos, rather than the Earth.

Copernicus was by no means the first person to suggest a heliocentric model of the heavens. His proposal drew significantly on earlier Greek and Islamic work and was argued primarily on the basis of reasoning rather than new evidence about the motion of the celestial objects. As a result, his work, which was not published until the year of his death, did not immediately lead to a fundamental shift in our understanding of the cosmos. It did enough however, to plant the seeds of doubt that would end up undermining the Ptolemaic system forever. Inspired by Copernicus, the Danish astronomer Tycho Brahe subsequently proposed a hybrid model in which the Moon and Sun moved around the Earth, but the planets moved around the Sun.

The real turning point came with the work of Kepler and Galileo in the first decade of the seventeenth century. First, Kepler swung a wrecking ball through the Platonic principles of the Ptolemaic model by abandoning the idea that orbits had to be made up of perfect circles. Instead, Kepler argued that the planets move on elliptical orbits around the Sun. Crucially, he was able to demonstrate that predictions of the position of the planets in the night sky using his theory of elliptical planetary motions were simpler and more accurate than those of the geocentric model, or indeed the first heliocentric model proposed by Copernicus.

Kepler's revolutionary ideas were swiftly backed up by publication of extensive observations of celestial objects made by the great Italian astronomer Galileo Galilei using a new type of scientific instrument - the astronomical telescope. Galileo was not the one who discovered that a combination of lenses arranged in the right way could generate a magnified image of a distant object (credit for which goes to the Dutch spectacle maker Hans Lippershey) but he adapted and improved the basic design to increase its magnifying power and turned it to the night sky.

The drawings and watercolours of his observations, rendered with a simple elegance that is elevated to a more sublime beauty by the significance of what they revealed, were the first rumours of a tidal wave that would sweep away the notion of the flawless, geometrically perfect, Earth-centred Universe completely.

Instead of the Moon being a perfect orb, Galileo saw the shadows of rough peaks and ridges that closely resembled the mountains of Earth. Instead of the Sun being a flawless flaming ball, he observed dark blemishes on its surface. Instead of the planet Venus being a single disc of light, he observed the planet going through a series of phases reminiscent of those of the Moon. Instead of every occupant of the sky appearing to move in perfect, sombre harmony, he noted four points of light near Jupiter that followed a regular dance around the planet in a manner that seemed to suggest they were orbiting it. Instead of the Milky Way being an insubstantial smoke, Galileo resolved it into a multitude of stars that were much fainter than those that could be observed with the unaided eye and which were certainly not neatly arranged in harmonious geometric patterns.

Still, the wave did not break immediately, but from this point onwards the downfall of the Ptolemy's geocentric understanding of the Universe was inevitable. As the seventeenth century drew to a close, Isaac Newton's theory of gravitation provided theoretical underpinning for the new model of planetary motion introduced by Kepler in a Universe that had lost its geometric aesthetic. The perfection of the heavens as envisaged by Plato, Aristotle and Ptolemy was, at least for the time being, consigned to history.

As use of the telescope for astronomy became widespread, it was not long before other observers began uncovering and cataloguing other objects in the night sky that had previously been hidden. Under the observing conditions of the times, unaffected as yet by the polluting glow of city lights, there would already have been awareness of celestial objects that are more exotic than the individual stars that lit up the sky. The great nebula in Orion would have been as familiar a sight for the workers of the land as for the educated elite; the famous double-cluster in Perseus would have appeared as a hazy smudge high up in the Winter sky, and even what we now know as the Andromeda galaxy - first recorded as a 'little cloud' by the great Persian scholar Abd Al-Rahman Al-

Sufi in the tenth century AD - would have been easy to spot if one knew where to look.

Now however, astronomers training their lenses on the sky began to discover other indistinct patches of light hiding in the gaps between the stars. On careful inspection, some of these could be resolved into clusters of individual stars but the limited power of early telescopes meant that the others seemed more diffuse, reminiscent of the prominent nebula in Orion but much smaller and fainter. A catch-all name of 'nebulae' began to be used for anything in the sky that fell into this category.

As more nebulae were discovered, they began to be included in catalogues of celestial objects, most famously those of the French astronomer Charles Messier. Messier was however far more interested in finding new comets than he was in nebulae and created his list primarily so that he had a record of objects that could otherwise be confused with cometary discoveries. The German-born William Herschel - who became famous for the discovery of Uranus - was also responsible, along with his sister Caroline, for methodically cataloguing thousands of nebulae from the end of the eighteenth century.

As it became clear that the night sky was littered with these irregular objects, the question of their nature became more important. Initially it seemed reasonable that, similar to the way in which the cloudy substance of the Milky Way could be resolved into a myriad of much dimmer stars, so too the diffuse nebulae were also just unresolved clusters. Over time, however, astronomers began studying the spectrum of light from nebulae to determine their chemical composition and were able to demonstrate that many contained considerable amounts of gas as well.

A century after the Herschel siblings began their work, the Earl of Rosse oversaw the construction of what was, at that time, the largest telescope in the world. The tube of the so-called

'Leviathan of Parsonstown', was so large that it had to be supported between two fixed walls on the Earl's extensive estate in Southern Ireland. This all but fixed the compass orientation of the telescope, although the elevation of the tube could be shifted pretty much all the way from one horizon to the other. At any particular moment therefore, the telescope could survey a single narrow strip of the night sky arcing overhead. To observe other objects, the Earl needed to wait patiently for the change in the seasons and the rotation of the night sky to bring them into view.

Despite its restricted pointing, the enhanced light collecting power of the Leviathan enabled Rosse to make more detailed observations of the shapes of the enigmatic diffuse nebulae. Famously, one of Rosse's sketches of the 51st object listed in Messier's catalogue showed that it appeared to have a distinctive spiral structure. It soon became apparent that many more nebulae also had a similar shape.

The discovery of spiral nebulae however made the question of their nature even more difficult and intriguing. While it seemed reasonable that irregularly shaped nebulae were similar in nature to the great nebula in Orion, the spiral nebulae seemed to form their own class of object. Still, however, by the start of the twentieth century many persisted in the belief that they were also groups of stars and gas similar to the more familiar nebulae found within, or in close proximity to, the band of the Milky Way.

There was however, a more radical alternative. In a characteristic leap of vision, the famous Enlightenment philosopher Immanuel Kant had popularised in the mid-eighteenth century the idea that spiral nebulae were 'island-Universes' that existed separate from our own Milky Way. Kant's conclusions built on the work of the lesser-known Thomas Wright, but were nevertheless remarkable for their foresight (unlike, sadly, his views on the superiority of Europeans over other races, which appear to have been more in keeping with their time).

The mystery about the nature of the spiral nebulae had transformed into a far more basic question about the way in which the Universe was arranged. If they had the same origins as the irregular nebulae, that supported a view in which the Milky Way was the only structure in the cosmos, to which all the objects of the night sky belonged. On the other hand, if Kant was right, then the stars that seemed to fill the heavens entirely were actually only part of a system that was one among many in the vast folds of the cosmos.

What was needed was a way to determine how far away the spiral nebulae were, so that astronomers could figure out whether they lay within the boundaries of the Milky Way or beyond it. At the time however, the only way to determine distances to objects beyond our solar system was by measuring their displacement over the course of several months due to an effect known as 'parallax'.

Imagine yourself in an art gallery somewhere in Italy. The room in which you find yourself is rectangular in shape with the walls painted a sophisticated burgundy. In the centre of the gallery stands a statue of Galileo, gazing up to the heavens with a sober expression of knowledge and wisdom; lining each of the walls are classical paintings of landscapes, ships at sea and various scenes of knights in battle.

Parallax is the cosmic equivalent of examining the statue of Galileo with your right eye closed, then noting how it appears to shift position relative to the paintings further away as you switch to viewing the room with your left eye closed instead. If you are standing close to the statue, it will appear to move much more than if you are standing by one of the walls of the gallery.

The experiment is an easy one to try out for yourself if you find yourself in a suitable exhibition (it works in other contexts too). It is also not too difficult to guess that, if you could measure the change in the statue's apparent position and the separation of your right and left eyes, then apply some maths, you would be able to determine its distance.

Astronomical parallax is pretty much the same thing, except replacing the right and left eyes with two locations on opposite sides of the Earth's orbit around the Sun. By precisely measuring the position of an object in the night sky from these locations (alternatively, on two occasions separated by half a year), and determining how much it has moved relative to the background of the rest of the night sky, it is possible to calculate the object's distance if you also know the diameter of the Earth's orbit.

Of course, there are a number of pretty thorny difficulties that faced astronomers in making accurate measurements of parallax. Furthermore, even with a telescope the vast majority of stars and nebulae showed no detectable shift over a six-month period at all. Indeed, this lack of observable parallax had previously been used as an argument against the heliocentric model. Either the Earth must be stationary in the centre of space, or the stars were located at unfathomably immense distances from us. While the principles of Plato still ruled, the idea that the sphere of the stars was located so much further away than the planets seemed a clumsy concept in a Universe founded on principles of mathematical perfection and order.

Even after the Ptolemaic system had been overthrown, its legacy must have continued to cast a shadow. It takes time for a new paradigm to be fully adopted. The demise of a theory does not mean that the ideas or ways of thinking that accompanied it are also lost immediately. As a result, the psychological step required to accept the notion of a Universe so many orders of magnitude larger than had previously been imagined must have been particularly challenging.

Slowly, however, the barrier was being undermined. Before Kepler and Galileo, the distance between the Earth and the Sun had been estimated by Ptolemy to be a little over 1,200 Earth radii. At that time the dimensions of the Earth had actually been measured fairly accurately, meaning that Ptolemy's measurements were equivalent to a distance somewhere in the range of 5.5 to 7.5 million kilometres in today's units. Kepler had realised that this figure was an underestimate, and subsequent measurements had boosted the distance by at least a factor of ten. By the start of the nineteenth century, the measurements had started to converge on the actual value of around 150 million kilometres.

The true scale of space was beginning to be revealed. As telescopes and observational techniques improved a handful of stars were also discovered that had a measurable movement across the sky due to parallax. In 1838 the first measurements of a star's parallax were made by Friedrich Bessel. The star in question - a fairly dim, reddish object located in the prominent Summer constellation of Cygnus - was found to be almost 100 trillion kilometres from Earth. This phenomenal distance was hundreds of thousands of times larger than that associated with the solar system.

Over the course of less than fifty years, the scale of the known Universe had ballooned by a factor of almost a billion. What was more, if the distance to Bessel's star was representative of how far our solar system's nearest stellar neighbours were, then the stars that showed no parallax shift must be much further away again. It was slowly becoming clear that the Milky Way was truly gigantic.

Still, Kant's island-Universe hypothesis suggested a cosmos in which even the Milky Way would be dwarfed by the rest of the Universe. If the Milky Way truly was one system among many and each of the other 'island-Universes' were of comparable size, they would have to lie at colossal distances to appear so small and faint.

Even by the standards being set at the time, this was a mind-blowing possibility. Understanding the nature of the spiral nebulae was the key to unlocking whether or not it was true. As the world entered the twentieth century however, astronomers were still not much closer to having a definite answer.

At long last however, a new discovery was made that offered a way forward. A class of stars called 'Cepheid variables' had been identified in the late eighteenth century that were characterised by regular, well-defined cycles in brightness. In 1908, Henrietta Swan Leavitt, who worked as an assistant at the Harvard College Observatory, reported the discovery of over 1,700 Cepheid variables in the Magellanic clouds - at the time known only as two particularly large and bright nebulous objects visible from the Southern hemisphere.

Intriguingly, Leavitt also examined a subset of the Cepheids in more detail and found that there seemed to be a relationship between how bright the stars appeared and the time it took for them to complete one complete cycle of variation. Leavitt also reasoned that because this group were located in the same cloud, they were probably also all situated at a similar distance from Earth. This mattered because the apparent brightness of an object depends on both its intrinsic brightness and how far away it is from us. Intrinsically bright stars located close to the centre of the Milky Way, for example, will usually end up appearing fainter than dim stars located in the same neighbourhood as our Sun, but brighter than tiny bits of rock in the outermost region of our solar system, which barely reflect any light at all.

As a result, the apparent brightness of a star usually cannot tell us anything immediately about its intrinsic properties. In the case of Leavitt's sample of Cepheid variables, however, her assumption that their distances were almost the same meant that differences in their apparent brightness must reflect differences in their intrinsic brightness as well.

Using the same logic, the relationship she had found between the apparent brightness of the Cepheids and the timescale by which they varied also pointed to a fundamental link between the duration of the variation and their intrinsic brightness.

This discovery may not immediately sound like much, but it promised to be game changing. The implication was that it should be possible to determine the intrinsic brightness of a Cepheid simply by measuring the time it took to go through one cycle of brightening and dimming. Once the star's intrinsic brightness was known, this could be compared to its apparent brightness in order to figure out its distance, and if the star was located in a spiral nebula then this would also act as a measure of how far away the nebula itself was.

A second paper in 1912, this time published under the name of the Observatory's long-serving director Edward Pickering, provided further evidence of the relationship. As an advocate for the educational advancement of women, Pickering appears to have been something of a rarity for his time and clearly acknowledged Leavitt's role in the drafting and results. As a result, credit for discovering the correlation - more formally known as the 'period-luminosity' relationship for Cepheids - has rightly been attributed to her.

All that was needed was to be able to calibrate the relationship in some way. It was not long before a number of Cepheids were found whose distances could be measured using parallax in order to calculate their intrinsic brightness directly. These provided a point of reference for the period-luminosity relationship, unlocked the methodology so it could be used for other Cepheids located across the night sky. The fog of uncertainty that had shrouded understanding of distances in space could at last begin to be lifted.

Sadly, neither Leavitt nor Pickering would live to see the full promise of the period-luminosity relationship fulfilled. A couple of years after his death in 1919, Pickering was replaced in his post by the first of the protagonists in the eventual Great Debate, Harlow Shapley. Shapley would go on to estimate the distance to the Magellanic Clouds using Cepheids to be in the region of 700,000 trillion kilometres. He also devoted much effort to measuring the size of the Milky Way, based on an alternative method that assumed giant associations of stars known as 'globular clusters' had a constant luminosity.

As a result, he believed the Milky Way to be several times larger than the distance he had measured to the Magellanic Clouds. The result seemed conclusive: the Magellanic Clouds - and in Shapley's view, probably the rest of the nebulae found in the night sky - must be located within the Milky Way.

This evidence was further supported by measurements of the apparent rotation of the spiral nebula M101 (the 101st entry in Messier's catalogue) by the Dutch astronomer Adriaan van Maanen, which were reported several years earlier in 1916. These suggested that if the nebula was located far beyond the Milky Way its outer edges would need to be rotating at a speed faster than that of light in order to match the observations. Einstein's theory of relativity had shown, however, that this was not physically possible.

There had also been observation of transient flashes of light, referred to as novae, in both the Milky Way and the spiral nebulae. At the time, it was assumed that these flashes came from similar events. If the spiral nebulae were distant, however, then the novae observed within them would have to be far more powerful than the novae observed in our galaxy, which seemed implausible.

There was, therefore, reasonable evidence to suggest that the island-Universe hypothesis was a flight of fantasy too far. The cosmos was huge, but not as vast as Kant had imagined.

On the opposite side of the argument however, evidence had also been quietly mounting that challenged the conventional view. Images of spiral nebulae taken at the Lick observatory in California, for example, had shown that they were even more abundant than had previously been realised and that spiral nebulae of substantially different apparent sizes yet remarkably similar form could often be viewed in the same region of sky. The idea that these spiral nebulae, located close to one another in the Milky Way, should form with such likeness but substantially different sizes seemed odd. It was much more reasonable that they were similar objects but located at vastly different distances.

The second of the protagonists, Heber Curtis, began working at Lick in 1902, where he had gradually become convinced that the spiral nebulae were located beyond the Milky Way. In 1910, he compared images of spiral nebulae taken over ten years earlier and found no evidence that the orientation of the spirals had changed. In contrast to the later evidence of van Maanen, this suggested that if the spirals were located within the Milky Way, then they must be rotating exceptionally slowly.

Just as Curtis drew different conclusions using the same type of evidence in the case of the rotation of spiral nebulae, he also used observations of novae to argue in favour of the island-Universe hypothesis. Curtis had noticed that isolated novae were seen more frequently in spiral nebulae than they were in the Milky Way. This could be explained if, in observing spiral nebulae, we were effectively sampling thousands of systems similar to our own. Even if novae were rare in individual island-Universes like the Milky Way they would be seen regularly across a large set of them.

Curtis also pointed out that it was far more common to find spiral nebulae at some distance away from the plane of the Milky Way, in contrast with the other types of nebulae that were concentrated close to or within it. The distribution was unusual if they originated in our own galaxy but made sense if they lay far beyond the Milky Way and could only be spotted through regions of the local star fields that were less densely populated.

The final weapon in Curtis' assault came from observations made in 1917 by Vesto Slipher which suggested that the spiral nebulae were moving away from Earth at tremendous rates. Put simply, the spiral nebulae were moving much faster than one would expect if they were gravitationally bound to the Milky Way, meaning they had to be independent objects.

Curtis became one of the best-known proponents of the island-Universe hypothesis, although his work was interrupted by America's late entry to the first world war. By the time the war was over however, the stage was set for science to debate its understanding of the nature of our Universe.

It is in this context that the academic jousting match between Shapley and Curtis became immortalized as 'The Great Debate'. In truth, neither side delivered the decisive blow that allowed for a clear-cut vision of the Universe to emerge victorious. Shapley is said to have not been a strong public speaker, and his principal concern in the lead up to the debate appears to have been avoiding any public embarrassment that might have undermined his application for a directorship at the Harvard University Observatory. As a result, the format of the event ended up being changed to two uninterrupted 40-minute presentations, rather than a back-and-forth discussion between the two astronomers.

Ultimately, the question was decided a few years later when Edwin Hubble observed Cepheid variables in the Andromeda nebula from the lofty vantage of the Mount Wilson Observatory, California, which overlooks Los Angeles. The results of his measurements were conclusive - the Andromeda nebula, and with it the rest of the spiral nebulae or 'galaxies' as they became known, lay far beyond the boundary of the Milky Way. Kant's intuitive leap several centuries prior had proven correct and with it the boundaries of the known Universe expanded once more.

Even though it did not provide the answer to the grand mystery of the spiral nebulae the Great Debate was as much a cultural steppingstone as it was a symbolic opportunity for the advancement of scientific knowledge. Today, it is relatively easy to take for granted the notion of a Universe that is made up of a multitude of galaxies scattered among the immense voids of space. We are taught this from an early age in school, and it is reinforced countless times in the news and through popular culture.

From our position of historic privilege therefore, it is hard to conceive what the impact on one's view of the world might have been for someone who lived in a period in which the Milky Way made up the entire structure of the Universe and other galaxies did not exist, or in earlier times still when the clockwork Universe of Ptolemy still dominated.

How, for example, would the subjective experience of seeing the Milky Way, witnessing an eclipse, or observing what we now know to be galaxies be affected? On clear, dark nights when no-one is around to see me behaving oddly, I have a ritual where I hang my head back so that my vision is completely filled by stars. A little while after my eyes have adjusted, there usually comes a moment when I feel a creeping sense of perspective, in which the field seems to gain depth and I become aware of the dizzying distances that lie between the Earth and the stars, and between those stars that are closest to us and the ones much further away.

In those moments, it often feels as though I am falling right into the galaxy from a height so unfathomable that I can barely notice myself moving. The sensation can be a little alarming to begin with, but this is quickly followed by a sense of quite blissful calm.

Whether this experience is best attributed to my awareness of the scale of space or the unnatural position of my head, it is tempting to say that humankind's vision is now much richer on account of us having a more detailed understanding of the Universe because this gives us even more reason to be moved by its scale and complexity.

While this is a common narrative, my suspicion is that there is something universal about humankind's wonder at the heavens. While its flavour may change, the essence of that wonder remains largely similar, regardless of how far our description of the Universe has developed. Astronomy has continued to fascinate and inspire throughout the millennia and there is no real reason, beyond the quiet dogmas of our own value system, to believe that those people who gazed at the stars hundreds or thousands of years ago felt anything less than we do now.

Questions about the equivalence of perspectives at different points of history may also have implications for how we view the essential work of science today. While the history of how our understanding of the Universe evolved up to the Great Debate demonstrates the tremendous success of science in developing an ever more detailed picture of our cosmos, there are also lessons within it that are worth reflecting on.

The first of these is the way in which previous generations of extremely clever and earnest scientists and scholars have been utterly convinced about the truth of a certain view of the Universe that has subsequently turned out to be fundamentally flawed.

The point here is not simply that those generations did not have access to crucial information about the innermost workings of nature but that they were convinced, on the basis of all the evidence that did exist, that other models of the Universe were wrong. When the geocentric model held power, for example, scholars would have been able to provide convincing and compelling evidence for why it was impossible for the Earth to be orbiting the Sun.

So deep has this conviction been that at various points there have been claims that the final answer to understanding how the Universe functions is just around the corner. None of these claims have subsequently borne fruit, yet the idea continues to persist, most recently in the form of Theories of Everything that seek to unite the two fundamental pillars of physics: quantum mechanics and general relativity.

It seems too great a conceit to claim that modern science is immune from being similarly blindsided. In some ways, the rapid success of modern science puts us at even greater risk of falling for such a conceit. So deep have we mined into the workings of the natural world that the theoretical frontiers of fundamental physics lie farther than ever beyond what we have been able to confirm experimentally. So faithful an ally has mathematics been in this adventure that it is now commonplace to hear that it represents the true nature of reality and should be trusted above and beyond what we observe of the physical Universe.

Strange as it might seem, we have returned to a place in which a large body of fundamental physics research is underpinned once more by a philosophy very similar to that of Plato's concepts of mathematical perfection.

None of which means that we should no longer trust scientists, particularly in the face of the trend towards post-truth populism in recent years that has been driven by the insidious influence of post-fact politics. We should, however, always remember to be humble in science, as we should in any field of academic thought, mindful as we are from the lessons of history that we are still capable of being wrong no matter how convinced we are of the truth of our convictions.

The flip side of this argument of course is that we should also remain open minded to all ideas, no matter how bizarre they might seem. Just as the heliocentric model would have been outdated or outlandish to someone convinced of the truth of the geocentric Universe, so the most bizarre-sounding proposals in physics cannot simply be disregarded because they seem fantastic.

In many ways, we are already very comfortable with this idea. Just five years after the Great Debate questioned the scale of the Universe, quantum mechanics revolutionised our view of nature on the tiniest of scales. With it came a level of oddness that had surpassed anything before. Today's understanding of the quantum world includes particles that do and do not exist; particles that interact with themselves, and an understanding of the nature of space that relies on the abstract concept of fields.

In our understanding of the Universe as a whole, science is also coming up with proposals that, on the face of things, would quickly be laughed away were they not backed by the combined power of an immense amount of intellectual effort and intuition. Whether we are talking about the idea that our cosmos is one of a huge multitude of other Universes separated in space, or that we constantly live mere millimetres away from parallel realities located in different dimensions, or the proposal that time loses its meaning at different stages of the Universe's evolution, there is a rich and beautiful landscape of bizarre and exciting ideas that science is already exploring.

Inevitably, some of these ideas will eventually be found to be no more than smoke and fantasy, but for one or two we may yet end up finding the faint, tell-tale echoes of their heartbeats in the true behaviour of reality. The scope of possibility in our future understanding of the Universe is enormous. Only time will tell what the subject of the next Great Debate will be, or the extent to which the Universe that emerges from it will bear any resemblance to the one we currently believe in.

Living with Solar Storms

It begins like a scene from an alien invasion B-movie. Imagine yourself in the operating room of an old-fashioned telegraph office somewhere in the dusty heartlands of North America. The date is the 28th August 1859, the Sun has set, and the evening has started drawing in. You are on your own in the office, accompanied only by the whirring and clicking of the telegraph machinery and the faint buzz of electricity.

The telegraph line has been working perfectly all day, but as the light outside begins to fade, there begins to be interference in the signal. Strange, as there were no lightning storms expected this evening, yet the signal is becoming impossible to discern. You reach to interrupt the power to the cables that transmit the telegraph signals, but as you do so bright sparks of electricity suddenly leap out from the apparatus, making you jump back and knock over your chair.

Slightly shaken, you back away cautiously from the table until you reach the office door, then out into the cool night. You hear nothing more from the disconnected receiver. The office door hangs open silently. Slowly, your breathing returns to normal and you begin to sense that there is something different about the night. The New Moon was due tonight and yet you can clearly see the bushes around you, as if the Sun had only just set.

It is then that you look up and see that the horizon is completely awash with light; a blood-red glow that fills that region of the sky and deepens to a rich crimson even as you watch, as though there is a great fire somewhere in the distance, the light from which is spilling into the sky.

Moments later however, the red wash is punctured by spikes of colour that streak up high overhead. In contrast to the red, these pillars of light are green, purple or a milky white. The sight is breathtakingly beautiful, and it is then that you remember having heard stories of such dancing lights before, but that they are usually only seen from the much colder realms further north.

Gradually your sense of foreboding is replaced with a rapt awe as the fantastic display continues for several hours before fading away to be replaced by the more familiar beauty of a moonless sky, resplendent with stars gazing down on Earth as if nothing had ever happened.

The events of the 28th August 1859 were unprecedented. Telegraph was a relatively new invention but had transformed communications across the industrialised world. Almost exactly one year earlier, the first trans-Atlantic telegraph cable had been laid, running between the picturesque Valentia Bay in County Kerry, Ireland and the Bay of Bulls Arm in Newfoundland, Canada. It was to be a revolution, unlocking rapid trans-continental communication between America and Britain, and with the rest of Europe.

On the night of the 28th however, telegraph networks across both the northern and southern hemispheres were disrupted to such an extent that they had to be shut down. At the same time, spectacular light displays were witnessed in the night sky across the globe. While the Aurora Borealis and Aurora Australis were familiar sights to those living within or near the chill bounds of the Arctic and Antarctic circles, they were rarely seen at more temperate latitudes, and never with such strength and vibrancy. US military ship logs even recorded the lights being witnessed from latitudes close to the equator. In the following days, the papers were filled with excited news of the unusual events. Extraordinary though they were, however, no-one could have predicted what would happen next.

Four days later, the widespread aurorae were back. Telegraph networks across the world again began to be disrupted. This time the fluctuations in transmission were even more extreme, producing showers of sparks from machinery that in some cases even caused paper left nearby to catch fire. As they did on the evening of the 28th, operators scrambled to shut off power to the telegraph lines, but then discovered something remarkable; even after the batteries that fed the system were disconnected, they could still transmit and receive messages through the network!

The telegraph system worked by sending pulses of electric current through cables connecting two locations. The pattern of the pulses encoded the message and could be deciphered at the other end of the line by an operator. In order to be transmitted however, there needed to be a current running through the cable in the first case, and this was usually supplied by battery packs connected to it.

The fact that the telegraph system continued to relay messages despite being disconnected could only mean that electric current was being generated in the wire itself. For almost two hours, operators on the East Coast of America could communicate with those on the West Coast without needing their own battery supply.

Meanwhile the fantastic heavenly light shows that had drawn such amazement only a few days earlier were even more intense.

Whereas the previous displays witnessed across North America had appeared predominantly in the northern horizon, the auroral glow now blanketed the entire sky, providing so much light that in many places it seemed as though the Sun was already rising. Accounts tell of people waking up believing it to be morning and starting their day; others who were able to read newspapers or print by the light of the sky, and of variations in the display so bright that they cast shadows on the ground.

There is even a well-known, if rather sad story of a small group of larks awakening and presumably beginning their morning calls before being shot by a local resident. The account of this incident in the New Orleans Daily Picayune several days later refers to the shooter as 'a gentleman', although I can see little that is gentlemanly about killing three small birds confused about the time of day.

As on the 28th August, these light shows were witnessed all over the world and must have made for a breath-taking sight from the ground. With hindsight, one can only imagine what they would have also looked like from space - the great orb of our planet, suspended amid the blackness; its otherwise dark night-time face enveloped by a suffuse blanket of feathery light constantly shifting its colour and form, as if it were alive.

The reaction to two such extraordinary events happening within such a short space of time was mixed. For some, the sublime beauty of the light shows served as a source of inspiration for poetry and art. For others their blood-red tones were seen as an ominous warning, heralding catastrophic events to come.

Within the scientific community however, they provided essential clues concerning the nature of the aurora themselves. Several popular theories of the day proposed terrestrial origins, ranging from the equivalent of high-altitude lightning to static produced by the rubbing together of ice crystals in the atmosphere, or the reflection of light from the polar ice caps.

The widespread and often quite detailed eye-witness accounts of the aurora from a large number of locations meant scientists could attempt to triangulate observations to determine the actual location of the auroral light. As a result, the lights were estimated to have extended up to 500 miles above the Earth's surface, which turned out to be roughly consistent with the definitive studies of auroral height made over eighty years later.

The height of the emission provided strong evidence that the driving force of the auroras was not terrestrial after all but came from a source beyond the atmosphere. The fact that telegraph lines had continued to operate even after being disconnected also demonstrated that whatever caused the aurora had something to do with electricity.

The question was, however, what was the extra-terrestrial driving force? It would be over one hundred years before the answer really began to take shape, but there were two sets of observations made in the lead-up to the second event that provided an important clue.

Richard Carrington was, in many ways, the archetypal Victorian gentleman astronomer. Born into comfortable surroundings, Carrington was sent by his father to study at Trinity College, Cambridge from the age of 18 with the apparent intention that he would end up joining the clergy. It soon became clear however that Carrington's interest in mathematics and mechanics meant he was better suited to a future in the physical sciences.

His interest in astronomy is said to have been triggered when he took a course on the subject delivered by James Challis, an English astronomer who - somewhat unfortunately given that he was, by all accounts, a perfectly capable director of the Cambridge Observatory - is best remembered for having unwittingly made a number of observations of the planet Neptune several months before its official discovery.

Carrington decided to become an astronomer and took up a post at the observatory of the University of Durham to gain experience of working in the field. The facility was still new, however, and lacked the equipment needed to deliver a programme of study that suited Carrington's ambitions. He held the post for a little over two years before retiring in March 1852 having been unable to acquire better instruments for the observatory. During his tenure he had, however, made good use of the University's well-furnished library and, it would seem, had determined that what he could not source from others he would just build for himself.

Carrington set up residence in Surrey and began building his own observatory. He went on to produce meticulous catalogues of stars that earned him recognition from the Royal Astronomical Society in 1857. While cataloguing the stars of the night sky earned Carrington widespread recognition, however, his most enduring legacy would be as a result of his systematic study of the Sun.

Between 1853 and 1861, Carrington made comprehensive drawings of the mysterious dark spots observed on its surface, called 'sunspots', which were to provide essential observational evidence in the discovery of the Sun's differential rotation (the fact that the Sun's speed of rotation changes with latitude), as well as for identifying the pattern of migration of sunspots during the Sun's regular eleven-year cycle of activity.

It was while Carrington was observing a group of sunspots on the morning of the 1st of September 1859 that he witnessed two intensely bright points of light emanate suddenly from the region. Carrington realised that he was witnessing something new and, after an initial moment of surprise, hurried off to find someone who would be able to corroborate the observation, only to find on his return several minutes later that the bright spots had all but disappeared. Unbeknown to Carrington however, there was another witness to the bright lights that day.

In the same year as Carrington set up residence in Reigate, former publisher Richard Hodgson was busy building his own observatory not so very far away. Hodgson had modified his telescope to allow him to observe the Sun from 1854 and happened to be monitoring the same region of the Sun as Carrington at the time the flashes took place.

Both men published accounts of their observations at the same time and in the same periodical a short time after. Unfortunately, however - as can often happen in the annals of science - the less pre-eminent co-discoverer tends to get forgotten in the story. As a result, the bright flashes on the Sun and the events that followed now tend to be remembered as the Carrington event.

At the time, no one in the astronomical community knew what the bright flashes were, and although he is often credited for having established a link between solar activity and the aurora, Carrington's own speculation on the matter was more reserved. In his report to the Royal Astronomical Society, he recognises the timing coincidence between the solar flashes and the global disturbances but is more cautious about linking them directly.

In fact, as we now know, there was indeed a connection between the two events. The flashes Carrington and Hodgson had witnessed were emitted from what may have been the largest solar flare ever recorded.

It is a prerequisite for any description of the visual appearance of the Sun that it includes a health-warning about the dangers of attempting to look at it directly. Despite it being around 150 million kilometres away from us, the Sun's light is so intense that even just observing it normally for a prolonged period without any form of telescopic aid can cause eye damage.

Any attempt to view a magnified image of the Sun has to rely on projecting its image onto a screen of some sort, to be studied indirectly, or else make use of extremely high-quality filters that block the majority of its harmful radiation (and, crucially, are resistant to scratches).

It is normal therefore for our first impression of the Sun to be that of a fairly constant, unchanging point of blazing light. Even from ancient times however, we have known that the Sun is not without its blemishes. Early evidence of knowledge about sunspots, for example, date back eight centuries before Christ, in the foundational Chinese text, the 'I Ching'.

Still, even sunspots also appear relatively tranquil at first glance; they do not tend to change shape or location quickly. The true dynamic nature of the Sun only begins to be revealed when the overpowering light of the surface is reduced or blotted out, either artificially or when the Moon passes directly between the Earth and the Sun, leading to a total solar eclipse.

In these moments, the much dimmer light emanating from the Sun's atmosphere is revealed. The result is mesmerizingly beautiful. Complex networks of wispy light stream away from its surface with an elegance that belies the violence that generates them. Occasionally, around the blocked rim of the Sun during a total eclipse you can also see small, pinkish blobs of light. These are 'prominences'; enormous loops of superheated gas that erupt from the Sun's surface and form giant fiery archways linking neighbouring sunspots. It is hard to conceptualise just how enormous prominences are. Suffice to say that the largest could contain the planet Jupiter - the biggest planet in our solar system - several times over.

Modern satellites and telescopes have built a picture of the solar environment that is as fascinating as it is terrifying. What appears as the surface of the Sun is in fact a layer of superhot plasma at a temperature of around 5,800 degrees Kelvin, which churns and buckles continuously as plumes of convective material rise up from its interior, releasing energy before sinking back into its fiery depths. Occasionally, horrifying tsunamis of solar material, that can be almost ten times as high as the Earth is wide, race across the Sun's surface at hundreds of kilometres a second. Completing the nightmarish picture are twisting streams of plasma forming hurricanes that whip about with terrible ferocity, and long ribbons of charged particles that flow continually into the Sun's atmosphere channelled along the lines of its magnetic fields.

In fact, most of the Sun's atmospheric weather is driven by magnetic fields bound into the convective material that lies below the surface, which become twisted and contorted as a result of its differential rotation. Sunspots originate as regions where these fields develop kinks that end up breaking out through the surface to form localised magnetic loops.

At the points where these leave the Sun's surface and return into it, the dredging of material from the solar interior is inhibited. This results in the area around these sites being slightly cooler than the surrounding material. The lower temperature also means that their brightness is less than neighbouring areas and it is this contrast that makes sunspots appear dark.

Solar flares are among the most extreme of solar weather. They form in the regions above sunspots where the looped magnetic fields can become stressed, contorting into arrangements that trap energy and the charged particles that are funnelled into them, preventing them from being channelled away. Under the right conditions, the magnetic fields become unstable and end up reconfiguring themselves. In doing so, they also unleash the energy that has been pent up within in a sudden explosive release.

These sudden bursts of energy are enormous. A typical flare has the explosive power around 500,000 times greater than the largest nuclear bomb ever detonated on Earth. The largest flares can be hundreds of thousands of times more powerful again. As a result, they provide an enormous outward kick to charged particles in the area, accelerating them close to the speed of light, which also ends up producing a lot of light - much of it as higher energy ultraviolet and X-rays, but also at radio frequencies.

Often, but not always, flares are also associated with even more disruption in the form of so-called 'coronal mass ejections', in which billions of tonnes of charged material are flung outward at high speed. These ejections are truly dramatic to watch - enormous bubbles that erupt from the site of the flare and out into space, expanding with fearful speed.

The material unleashed by these solar storms can be travelling thousands of kilometres every second as they race away from the Sun and plough their way through space. This means they can cover the distance between the Sun and Earth in just a day or sometimes even less. If the Earth is standing in the way at the time, there is nothing that can be done to prevent us from being struck.

Perhaps one of the most underappreciated characteristics of the Earth that make it hospitable to life is its relatively strong magnetic field, which arises primarily as a result of flows of liquid metal in its outer core. The material in these flows is not intrinsically magnetic - conditions near the centre of the Earth would not normally support the formation of long-lasting magnets - but their continuous movement generates electric currents that, in turn, also induce a magnetic field. Every time we use a compass, therefore, we are making direct use of the fact that the Earth's deep interior is in constant motion.

The Earth's magnetic field is not just a means of helping us navigate the globe though. Its influence extends out into space and forms a sort of protective magnetic cocoon around our planet which we call the magnetosphere. As we have seen however, the

Sun also has its own magnetic field which, outside periods when it is experiencing a solar storm, is manifest in a constant flow of material from the Sun, also known as the 'solar wind'.

The solar wind acts to continuously shape the magnetosphere, moulding it like the smoothing strokes of a sculptor's palette. As a result, the magnetosphere gets constantly squashed and extended as the Earth rotates. On the planet's daytime side, its magnetic influence extends around 120,000 kilometres into space, or about a third of the distance between the Earth and the Moon, but on the night-time side its reach is over ten times further.

The primary importance of the magnetosphere to life on Earth is that it shields the planet from many of the harmful ionising particles emitted by the Sun, which could otherwise damage the DNA of cells, by channelling them around and away from us. This not only protects life directly, but also slows the rate at which we lose gas from the atmosphere as a result of it being effectively dragged away by the solar wind. By comparison, Mars does not have a strong magnetic field and loses around ten times more of its already meagre atmosphere to the stripping effect of the solar wind than the Earth.

The magnetosphere is not, however, completely impenetrable. It has two areas of weakness, one that allows solar particles in from the daytime side above the North and South Poles, and a second which lies far beyond the Earth on its night-time side, at a location where the solar wind's magnetic fields reconnect, having been effectively separated by their first encounter with the Earth's magnetosphere. This second site acts as a back door through which charged particles in the solar wind slingshot back to the planet from the night-time side and are funnelled in towards its polar regions.

These flows of charged particles are ultimately what cause the auroral lights. They do so by generating electric potentials around the Earth that then accelerate particles - either from the solar wind itself or from an existing pool of low energy charged particles trapped in the magnetosphere close to Earth - into the upper atmosphere. Here they collide with neutral particles in the atmosphere, which transfers energy to them.

The excited atmospheric particles do not however, tend to hold on to their energy for long, and end up settling back into their original state by releasing the energy they have gained in the form of light. The exact colour of the light dependents on what type atom is excited - the red glow observed in the Carrington event was due to oxygen atoms and is actually reasonably unusual to see on its own, as these transitions are usually swamped by more frequent collisions that take place with different atoms at lower altitudes, giving aurora their more common green glow.

Aurorae are, therefore, a visible manifestation of the invisible, dynamic link between the Earth and the Sun, and are generally visible on Earth from latitudes of about 60 to 75 degrees north or south. During periods of high solar activity their intensity can increase substantially as the magnetosphere takes the strain of the additional solar weather.

If the Earth is hit by the coronal mass ejection associated with a particularly powerful solar storm, however, the magnetosphere's deformation becomes even more exaggerated. This allows electric potentials to develop at much lower latitudes, leading to auroral displays, and reduces the capacity of the magnetosphere to deflect high energy particles.

In itself, this may not sound particularly scary, but the Carrington event demonstrated that powerful flares and coronal mass ejections can also have a real effect on the infrastructure of society itself. Estimates made of the cost of disruption worldwide to the telegraph network during the Carrington tend to fall somewhere around $300,000 at the time, which equates to around $9.5 million in today's money.

In the grand scheme of global spending, this amount of money is barely noticeable and not particularly worth getting wound up about. The comparison completely underestimates, however, the threat to today's society from the effects of solar storms.

Over the past 150 years or so since the Carrington event, our vulnerability to solar weather has increased exponentially. What is worse, there is relatively little we can do to protect ourselves from harm in the case of extreme events.

The fallout from solar storms typically come in two waves. First, the Earth is bombarded by the high energy radiation and relativistic particles generated by the flare itself. The radiation arrives first within 10 minutes of the eruption, followed shortly after by the particles. Both pack enough punch to knock electrons out from atoms in the upper atmosphere, or ionosphere, and in so doing alter the transmission and reflective properties of the atmosphere to radio waves.

This is a problem because the world depends heavily on a multitude of satellites in space to function normally. These control everything from global telecommunications to credit card payments and military networks. In particular, the global positioning system relies on a fleet of 27 satellites and underlies not only our ability to find the fastest route to the nearest coffee shop and locate new Pokémon characters in the centre of Liverpool, but also the functioning of financial transactions across the world and the protocols that keep the internet running. All these satellites rely on radio communications with the Earth in order to relay information properly, and if these were interrupted as a result of interference from a solar storm, the impact would be felt quickly and could end up being profound.

Satellites also face more direct threat from the relativistic particles, which can penetrate the outer layers of the spacecraft and cause discharges in its circuitry, leading to damage and sometimes even the total failure of the satellite.

Problems with radio communications also pose serious issues for airlines whose planes rely on radio to relate their position and altitude to ground controllers. Interference in these updates therefore undermines the careful control that is kept over the enormous number of flights that take place every day.

After the overture of the flare, however, the second wave of effects comes as the coronal mass ejection itself slams into the Earth's magnetic fields like a cosmic juggernaut. Once more, satellites are first in the line of fire. The electrical surges that ended up enabling telegraph communications to continue during the Carrington event can also occur in satellites, overloading their electronics and causing them to fail.

Electric surges pose threats to Earth-based systems as well, in particular the large transformers that are key to the distribution of domestic electricity. Currents induced in overhead power cables can lead to excess heating in the transformers, which can cause long-term damage or failure, as well as instability and potential collapse of the power network, either of which could result in the loss of electricity to entire regions. In the worst cases, the repair of damage to the system could take a long time to recover.

It is often tempting to wave away the harbingers of doom who point their trembling fingers variously at solar storms, asteroids, or super-volcanoes on Earth and promise destruction. While it is true that we should not be overcome by fear of events which we have little control over, it is also clear that the warnings about solar storms are not just idle threats. The US government takes the risk seriously enough that it provides online advice to citizens on steps to take in preparation for the disruption that could be caused, which reads something like a survival plan for an apocalypse.

There have also already been a number of close calls with solar storms. When communications with a US satellite system that provided early warning of a nuclear missile launch began to experience interference at the heart of the Cold War in May 1967, for instance, one explanation was that the defences were being deliberately interfered with. The fear was that the Soviets wanted to take down the system in order to prevent America from detecting the launch of an all-out strike. Clearly, if this were correct, it could have had serious military consequences. US military officials scrambled to try to figure out whether the jamming was deliberate or not.

Luckily for them - and perhaps for the world as a whole - the Environmental Science Services Administration and the US Air Weather Service had detected flares erupting on the Sun prior to the interference and were able to link these to the issues experienced communicating with the satellites.

Calamity was avoided, but only by a narrow margin. The Environmental Science Services Administration had been set up just two years earlier to study the Earth's meteorology, and even though the Air Weather Service issued alerts directly to the US Air Defence Command, NORAD, other crucial agencies were oblivious of the effects of space weather and did not take them into consideration in their decisions.

The unit responsible for the Air Weather Service's space alerts was also only co-located with NORAD to provide a round-the-clock space weather monitoring service in 1965. Had the flare taken place a few years earlier, who knows what the outcome might have been.

The near miss of 1967 encouraged further priority to be given to monitoring and understanding the potentially serious impacts of space weather.

Despite this, we have suffered the cost of finding ourselves caught in the line of fire of the Sun's outbursts on more than one occasion. The famous solar storm of 1989 led to power failures across Quebec, cutting electricity to millions of people for around nine hours; the evocatively named Halloween solar storm of 2003 caused outages in Sweden, loss of radar contact with planes, and damage to a number of satellites, and a series of solar flares in 2017 led to a temporary loss of radio communications during the emergency response to hurricanes Irma, Katia and Jose in the Caribbean. The storm that has concerned scientists the most, however, was one that did not cause any damage at all.

In the Summer of 2012, the Sun was midway towards the peak of its regular cycle. From the start of July, astronomers had observed a new sunspot region that, over the course of the next few weeks, produced no less than seven moderate flares and one major flare, as well as two coronal mass ejections.

Then, on the 23rd July, the region emitted an exceptionally strong storm, which included two coronal mass ejections in quick succession. The storms earlier in the month had effectively acted as a snowplough, clearing a path through the inner solar system's interplanetary matter to make way for this final eruption. As a result, the material ejected from the Sun retained much of its speed and ferocity as it careened through Earth's orbit around seventeen hours later.

Luckily for us, it swept through a locality of space that our planet had left just a week or so before. The calm mathematics of the Earth's orbit had saved us from a direct hit, sweeping us out of harm's way just in time as the solar material stampeded past. Had we not been so fortunate, however, researchers estimate that the resulting storm on Earth could have been at least as powerful as the Carrington event, and possibly even more so.

It is hard to determine precisely the extent to which satellites would have been affected or how power systems would have coped to determine the impact such an event would have had. So much depends upon exactly when and where the storm would have struck. If the conditions were right to maximise the impact of the storm, however, it is estimated that the financial cost of recovery could have run into some trillions of dollars and take many years to complete. Whatever the figure, it is clear that the Earth had narrowly avoided major disruption.

There will come a time, however, when we will not be so lucky. Carrington type events are estimated to occur roughly once every century, but there is no hard and fast rule that the Sun will respect, and the likelihood of another such major event affecting us within the next few decades is uncomfortably high.

As things stand, our only protection from solar storms is attempting to forecast in advance when they might occur and how damaging they might be. At least with some warning, power grids could be switched off temporarily to avoid the effect of power surges and steps could be taken to protect satellites by switching them into safe modes. This would not be enough to prevent damage in the event of a major storm, but it would give us more reason to cross our fingers and hope for the best.

Today, the National Oceanic and Atmospheric Administration forecasts space weather around the Earth based on observations of the Sun and provides live updates and warnings of disruption. These are, however, largely reactive, based on how much activity is observed, or following the detection of a flare or coronal mass ejection.

As a result, we currently only have a limited window of opportunity to prepare. What we would like is to be able to reliably predict when storms will occur in advance of them happening. Traditionally, such predictions relied on identifying apparent relationships between the characteristics of sunspots regions and the likelihood of a coronal mass ejection. Unfortunately, these approaches lack a strong underlying physical mechanism to explain why the relationships might be reliable, which undermines confidence in their general application, and suffer from significant errors. In 2020 however scientists moved one step closer to producing a predictive system that relies on an understanding of the physics of active sites instead.

The Institute of Space-Earth Environmental Research at Nagoya University in Japan is a leading interdisciplinary institution whose research aims to understand the complex interdependencies between the Earth, Sun and space environments. Established in late 2015, its philosophy is to treat the interactions between these entities in a holistic manner, to better understand not only how the dynamics of the entire system influences the environment on Earth, but also to explore issues around the expansion of human activity into space.

In July 2020 a group led by the Institute's Director of research, Kanya Kusano, reported the results of using a new approach for predicting when major solar flares would occur, referred to as the 'kappa scheme'.

The kappa scheme relies on a model of how solar flares are triggered based on the formation of so-called 'double-arch instabilities' in the magnetic fields surrounding sunspots. These instabilities occur when two magnetic loops form close to one another, such that the descending portion of one arch is found near to the ascending portion of the other. Under the right circumstances, the magnetic field lines that make up these sections of the arches become twisted, and when they move within a critical distance of one another they can suddenly connect, forming a shallow bridge between them.

The central region of the resulting 'double-arch' structure is unstable and will tend to move outward from the surface of the Sun. The movement leads to a cascade of similar connections higher up in the magnetic field, increasing the instability still further, until there is a sudden release of the energy contained in the magnetic field in the form of a flare. If this model is correct, the degree of twisting in the magnetic field lines is an indicator of the likelihood that a double-arch instability will form in the first place. This is the first element of the kappa scheme.

The second element is the proximity of instabilities to features of sunspot regions known as 'polarity inversion lines', which represent the boundary between areas on the solar surface where magnetic fields switch their direction. These have long been known to be associated with flares, but the underlying model of the kappa scheme suggests this is because the large gradient in the field allows double-arch instabilities located close to inversion lines to release more energy than those that form further away.

Kusano and his team applied the kappa scheme retrospectively to observations of around two hundred sunspot regions from eleven years of solar observations between 2008 and 2019. They were able to successfully predict seven of the nine most powerful flares that took place on the Earth-facing the side of the Sun during that period around twenty hours before the flares took place.

Not only that, but the scheme was also able to pinpoint the location of the flare and provide estimates of how powerful it would be. Integrating this kind of detailed information into solar weather alerts would be a significant step forward in forecasting where and when the most powerful flares are likely to take place, and how disruptive they may be.

Being able to accurately forecast when solar flares will take place is, however, only one part of the story. While the high energy radiation and particles from a flare will travel more or less in a straight line through the intervening space between the Sun and Earth, the passage of coronal mass ejections is more complex. The initial direction and speed of the ejection can vary substantially and its shape and motion through space depend in part on the state of the solar wind through which it travels. Each of these factors can influence whether or not the storm hits the Earth in the first place and, if it does, its strength when it arrives.

Anticipating the effect of solar storms on the Earth therefore also requires some way of predicting the passage of a coronal mass ejection through space. In September 2020 a group from the University of Reading, led by Dr Luke Barnard, published details of a new approach to predicting when coronal mass ejections will hit the Earth, based in part on analysis carried out as part of a citizen science project called 'Solar Stormwatch'.

Citizen science projects do what they say on the tin by enlisting volunteers from the general public to participate in scientific research. Their popularity over the past few decades has increased substantially as a result of the internet, and there are now thousands of active projects taking place across a wide range of fields that anyone can get involved in. The manner in which the volunteer is involved in these projects varies significantly, from generating data by participating in games or keeping records of the number of birds seen in a certain locality, to offering the use of their computer's processing power when it is otherwise standing idle.

Citizen scientists involved in the Solar Stormwatch project, however, took a more active role in the analysis of data collected by two satellites called STEREO-A and STEREO-B. The spacecraft flank the Earth like tiny sentinels in its orbit around the Sun - one ahead, the other behind - and carry instruments designed to image outflows of material from the Sun and monitor space weather.

Citizen scientists had two primary roles in the project. The first was to examine movies of coronal mass ejections and identify the times at which storms entered the view of each spacecraft, as well as monitoring how they then moved through the field. Comparing data for both satellites allowed for estimates to be made of the ejected material's initial speed and direction of travel.

The second role was to track the progress of the coronal mass ejection out as far as possible as it left the vicinity of the Sun. This allowed for an early profile of its route through space to be established before the leading edge of the ejected material became too dispersed to identify.

Since its launch in 2010, the Solar Stormwatch project has characterised over a thousand coronal mass ejections using this approach. The team's next step was to see if they could use that data to improve forecasts of the arrival time of coronal mass ejections at the Earth.

The problem is that understanding precisely how a hot, energetic bubble of charged particles interacts with the interplanetary magnetic field and solar wind particles that lie between it and the Earth is extremely complex. Existing forecasting methods use various simplifications to estimate the arrival times of coronal mass ejections, but these can lead to large errors in the predictions. Remove these shortcuts, however, and the models become too complex and unwieldy to use for live forecasting.

Barnard's team took a different approach. Looking at four storms that took place in 2012, they simplified the underlying model used to track their propagation through space and ran these models hundreds of times instead. During each run, they used slightly different conditions to see how the prediction for the arrival time at Earth's orbit changed depending on the likely properties of the storm itself and the space through which it was passing. The results could then be collected together to provide a range of possible arrival times for the storm on Earth and tested against their actual timings.

Crucially, however, the team used a comparison between the model results and the initial profiles of the storms collected through the Solar Stormwatch project to refine their calculation of the storm arrival times. This additional intelligence about the storms was to prove crucial. The difference between the modelled and actual arrival times reduced on average by about twenty percent, while the uncertainties in the modelled results were also cut by a similar amount.

Both the kappa scheme and the approach to predicting the arrival times of coronal mass ejections still need further work. The kappa scheme missed two major flares in the data set it was tested on and predicted a number of flares that did not subsequently occur. It is reassuring to note, however, that the missed flares were unusual inasmuch as they were not also accompanied by coronal mass ejections, so these events represented substantially less threat to the Earth.

As for the false predictions, while we should be less concerned about storms that do not occur than the ones that do, we also would not want a system that cries wolf too many times. Clearly, therefore, there is still work needed to improve the model before it can be used reliably for forecasting.

Similarly, the work of the Reading team on the arrival times of coronal mass ejections at Earth was intended more as a proof of concept to demonstrate the feasibility of the method, rather than a final prediction system. The team continues to refine their approach and are advocating to make sure that future solar satellite missions include instruments that will produce the kind of data used in the Solar Stormwatch project.

Regardless however, it seems as though we are slowly making progress in our ability to predict when solar storms will hit the Earth and what effect they will have when they do so. Perhaps one day we might even be able to watch weather forecasts that include coverage of solar activity as regularly and with as much confidence as we do our regular weather forecasts today.

Until then however, much of the towering stack of technology that underpins the functioning of our society remains vulnerable. As our dependency on this technology becomes ever greater, so do the risks to our modern way of life and the costs of recovery from a storm as extreme as the Carrington event.

Living in close proximity to the nuclear furnace of a star is not easy. The irony is, of course, that without our stellar guardian, life on Earth would simply not be possible. It is a further reminder of just how fragile our existence on this planet really is, and how vulnerable our way of life remains when faced with the raw power of the Sun.

The Path of No Return

I f you had been standing outside somewhere in the south-west region of the UK shortly after 10pm on the 30th May 2020 and happened to be looking into the darkening skies of the west, you might have noticed what looked like a relatively bright star, moving purposefully across the sky.

Without knowing it in advance, there would have been relatively little about this shard of light to suggest that it was mankind's only permanent outpost in space, the International Space Station (ISS), completing another of its 90-minute orbits of the Earth. From the lofty vantage of its altitude around 400 kilometres above us, the space station was reflecting the last bright rays of the Sun that had already set from the perspective land below, as it swept ghostlike along its path around the planet.

If you know what to look for, it is relatively easy to spot satellites at night. They often appear like fairly dim stars, moving quietly and unobtrusively among the field of real stars, as if they know they are imposters. Sometimes these dots seem to disappear mid-flight, blinking out of existence as they pass abruptly into the Earth's shadow and out of sight of the Sun. The ISS is, however, the only one of Earth's vast multitude of artificial satellites where you can make out its shape fairly easily with a reasonably sized amateur telescope and the right tracking. The station's structure comprises of a long spindly central beam from which its solar panels extend like the fronds of a fern, making it easy to distinguish. With a little more telescope-power, it has even been possible for backyard astronomers to catch sight of the fuzzy forms of astronauts as they take part in spacewalks around it.

On this particular night, however, the ISS had a tail. A few minutes after it had passed overhead, another point of light appeared over the horizon moving in the same direction, as if it were secretly tracking the space station.

Almost two hours earlier, the sleek black and white tube of a Falcon-9 rocket had stood balanced upright on its launch pad at the Kennedy Space Centre in Florida, wrapped loosely in brooding clouds of wispy white vapour. Perched at the top of the rocket at a height of about sixteen stories, overlooking the thin sandy beaches that define this stretch of the Florida coastline, was the squat white dome of a Crew Dragon capsule.

Inside the capsule were two astronauts, Bob Behnken and Doug Hurley. They were about to embark on a nineteen-hour flight to the ISS. Strapped into thin moulded chairs with the same black and white colour combination as the rocket, they occasionally tapped a set of screens located less than a foot from their helmets as the final flight checks were completed, then settled back and prepared themselves for the shuddering violence of launch.

Moments later, the sharp hiss of pre-ignition gave way to the thunderous roar of hot gases being funnelled out from the base of the rocket, thrusting it up into the sky. The Falcon-9 rose smoothly past the launch pad's support mount, balanced precariously on a pillar of flame. In just thirty seconds, it had accelerated to a speed of 460 kilometres per hour and was almost two kilometres above the ground. Two minutes and forty seconds after launch, the rocket was little more than a speck in the sky. By this time the fuel in the first stage of the rocket, comprising about half its total height, had been exhausted and the empty funnel detached from the rocket's upper section.

For the next five minutes, the second stage of the rocket took over, powering itself and the Crew Dragon to an altitude of 200 kilometres and a speed of 27,000 kilometres per hour. Eventually its fuel was also spent and, after careful checks on the rotation of the two ships, the second stage also separated, its hollow shell drifting away into space, leaving the Crew Dragon to continue on its own.

The Crew Dragon made its first pass over the UK around twenty minutes after launch, but at this point the Sun had not yet set and the gleam from the spacecraft was difficult to make out against the dwindling brightness of the sky. One hour and forty minutes later, however, the capsule had already travelled around the Earth and the skies were dark enough to witness the high-altitude game of cat and mouse between it and the ISS.

It had been a very American launch. Trips by astronauts to the ISS do not normally make the national headlines but, ever since the space shuttle programme had been retired in 2011, NASA had relied on Russian Soyuz rocket and spacecraft, launched from the dusty launchpad of the Baikonur cosmodrome in Kazakhstan, to blast its astronauts into space. The Crew Dragon was different - it was the first American-built, American-launched craft to be used to ferry humans into space for almost a decade.

The long history of political tension between America and Russia is of course well known, as is the symbolic and strategic role that space exploration played in the early stages of the Cold War. The two nations effectively called a truce to the space race in 1975, when the last of the Apollo missions docked with a Soyuz spacecraft and American astronaut Tom Stafford reached over the threshold of the adjoining hatch to shake hands with his Russian counterpart, the famous Alexei Leonov.

Although the spirit of peace and cooperation embodied by that mission was undermined just a few years later as relations between the countries once again deteriorated during the Soviet-Afghan war, they remained uneasy neighbours in space. In the early 1990's, American President George Bush Senior signed an agreement with Russia's first president Boris Yeltsin that paved the way for greater cooperation in space exploration. Not long after, the American shuttle Atlantis docked with the Russian space station Mir, and crew from both nations once more shared a home together in space.

By the end of the millennium, the relationship between America and Russia in space was enjoying something of a golden age, with the two nations joint collaborators alongside Japan, Europe and Canada in the ISS. When the cost of running the space shuttle fleet began to grow and the programme was shut down, Russia rockets provided a lifeline for the American manned space programme.

It takes longer however for the political and cultural memory of decades of tension to disappear. Despite the accord, America's dependence on Russian space vehicles could never have sat comfortably with everyone. Worse was to come when NASA's Constellation project - which was to act in part as a replacement for the shuttle - was also scrapped in 2010, again as a result of escalating costs.

America's reputation for delivering affordable manned spaceflight was suffering and the scientific programme for the ISS was being propped up by the consistent performance of Soyuz. Politics became an issue once more in 2014, when Russia annexed the Crimean Peninsula, which juts out into the Black Sea off the southern coastline of Ukraine, following protests and civil unrest concerning the country's alignment with Russia that resulted in the pro-Russian Ukrainian president Viktor Yanukovych fleeing the capital Kyiv.

In response to the annexation, Western sanctions were placed on high profile Russian individuals. Amid the posturing that followed Russia's deputy Prime Minister Dmitry Rogozin - who was later to become the head of Russian space agency - famously tweeted that the US should deliver its astronauts to the ISS using a trampoline, rather than relying on Soyuz.

NASA needed a new approach. Ever since the start of the Obama administration, they had been working with aerospace companies to develop plans for commercial spacecraft and had already used privately manufactured launch vehicles to fly payloads into space. Now the agency shifted the full focus of its attention on producing the next generation of American spacecraft to the private sector as well.

In September 2014, NASA announced the award of contracts worth over 6.5 billion dollars to aerospace giant Boeing and the much smaller SpaceX company, which was founded by the charismatic South African entrepreneur Elon Musk shortly before he became famous as the head of electric car manufacturer Tesla. The contracts were intended to develop spacecraft capable of transporting humans to the ISS.

The race was on for Boeing and SpaceX to develop their designs and demonstrate that they could successfully launch unmanned versions of their spacecraft and dock with the ISS before they would be allowed to carry astronauts into space.

Technical issues dogged both sides, but in March 2019, SpaceX became the first to send an unmanned version of its Crew Dragon capsule to dock with the ISS and return to Earth. Despite the same capsule exploding just a month later during systems tests, the door was open for them to attempt a flight with human crew next. Meanwhile, the Boeing team was experiencing bigger problems and after its Starliner capsule failed to dock with the ISS in December 2019, it was clear SpaceX would have the first shot at delivering astronauts to the space station.

Having overcome every hurdle to reach this point, however, the stakes for SpaceX and for the American programme of human spaceflight were now at their highest. Failure of the mission at this stage would be a disaster: for the astronauts whose safety relied on everything working correctly; for SpaceX whose vision depended so much on its success, and for NASA who could not afford another failure.

Little wonder therefore that the media hype around the mission was huge. Two days earlier during the first attempted launch, in the midst of a country battered by the Covid-19 pandemic and widespread protests over the death of George Floyd, Donald Trump and vice-president Mike Pence travelled to Florida to witness lift-off. The launch had even been touted as an opportunity to unify America, just as the Apollo missions had done several decades earlier.

As if acknowledging that long-standing inequality and a virus for which humankind had no resistance could not be forgotten so lightly however, the normally reliable Florida weather had other ideas. The launch was called off sixteen minutes before lift-off due to high winds.

Second time around the weather held. Social distancing measures meant there were no large crowds present, but the online coverage, boldly branded as 'Launch America', was watched by many millions.

The coverage was packed with patriotism. As the Falcon-9 began its fiery ascent, commentators announced a new era for American spaceflight; excited anchor-men and women proclaimed it a historic moment, and panel experts were left speechless with emotion. "Launching American astronauts on an

American rocket, from American soil" became the catchphrase of the evening with large flags on prominent and frequent display.

The display was, no doubt, intended primarily for a domestic audience, but there was no avoiding the international element as well. Much was made of Elon Musk's joking comment immediately after the launch that "the trampoline is working" - a belated comeback to Rogozin's 2014 barb - but Musk, for whom this moment really did represent the culmination of an objective towards which he had been working for many years, was also more generous in his tone, claiming the moment not only for America but for anyone who shares the dream of exploration.

Beyond the headlines however, the launch of the Crew Dragon was the herald of a much quieter but far more significant revolution. Since Yuri Gagarin became the first person to be sent into space and orbit the Earth, national agencies have been the gateway to space exploration. In 2004, however, the US congress passed the Commercialised Space Launch Amendments Act, with the aim of promoting the development of the human space flight industry. For the first time, private entities could legally launch humans into space.

The cost of space flight however remained extremely high, but with increasing competition between the likes of Boeing, SpaceX, Virgin and others, as well as the pioneering of reusable launch rockets, the costs of hauling stuff into space has gradually begun to fall over the past ten years or so.

The successful launch of astronauts to the ISS opens the door therefore to a new era of spaceflight. The theory is that a flourishing private industry of manned launches will drive down the costs of getting humans to and from Earth orbit, leaving national agencies to pursue space missions with much grander objectives, such as NASA's Artemis programme, which aims to return humans (including the first woman) to the Moon and begin establishing a more permanent human presence in lunar orbit. In the slightly longer term, the aim would be for the Moon to act as a steppingstone for trips to Mars.

The question is, of course, who will get to the red planet first? It is no secret that Elon Musk's dreams go much bigger than simply shuttling astronauts to the ISS. His aim of sending astronauts to Mars and developing a colony there has been the driving vision behind the phenomenal success of SpaceX since its beginning. At the time of writing, the background to his Twitter feed even depicts the red planet slowly transforming into Earth (at least if you read the graphic from left to right - the opposite way is somewhat less encouraging), while the Mars page on the SpaceX website sets out a vision of humans growing plants on the planet and slowly adapting its environment.

Musk set out his vision even more clearly at a virtual summit to discuss humans on Mars in September 2020, in which he outlined his aim to have established a self-sustaining city on the planet by 2050, which could reach a population of around a million people. To achieve this would require an enormous fleet of spacecraft to shuttle settlers, materials and equipment across more than 100 million kilometres of interplanetary space.

Whether or not Musk's long-term plans can ultimately be realised, the ambition of SpaceX launching a mission to Mars is already well on its way to being delivered. In August and September 2020, the company carried out two success tests of the prototype for its Starship spacecraft. The design of this monster spaceship looks like something from a science-fiction movie. The intention is that it should eventually be capable of transporting up to a hundred astronauts at a time to Mars.

Footage from the early tests show what appears to be a giant metallic silo container - indistinguishable at first sight from the structures of the launch site - rising slowly 150 metres into the air on a bed of exhaust gases before slowly descending back to the

Earth and landing upright once more. By the end of the year, the Starship had also carried out its first high altitude flight test.

Musk's timeline for getting to Mars is ambitious. As things stand, he plans to launch a crew bound for Mars around the same time that NASA is aiming to be landing once more on the Moon. The progress, however, is impressive and at this point seems unstoppable, as if carried on the wave of Musk's unwavering determination and vision. In the face of SpaceX's success so far, it seems possible that his objective may just about end up being achievable.

Of course, getting to Mars is one thing, but being able to sustain a presence there is something quite different. One way or another - and as Musk has reiterated on numerous occasions - there is a good chance that the journey of the first settlers to Mars will end up being only one-way.

While the dream of exploration can easily become an obsession that inspires us to push the boundaries of possibility to their farthest limits, it is not the only reason that private companies are increasingly becoming interested in space. Since American investment management magnate Dennis Tito became the first private citizen to pay for the privilege of flying into space and visiting the ISS, the potential of the space tourism industry has slowly but surely been growing.

In the lead up to the Crew Dragon launch, two companies - Axiom Space and Space Adventures - announced they had organised deals with SpaceX to fly tourists into space. Axiom will deliver its passengers for a ten day stay on the ISS, while the Space Adventure tour will be a free-flying trip that will fly higher than the orbit of the ISS.

The announcement followed the much-publicised unveiling in 2018 of Japanese billionaire Yusaku Maezawa as the mysterious first customer for a SpaceX trip around the Moon, currently scheduled for 2023. In time, Axiom is planning to have its own module added to the structure of the ISS in order to house tourists - the ultimate in exclusive hotel accommodation, although the amenities will be somewhat different to its five and six-star cousins back on Earth.

With a price tag believed to run into the tens of millions of dollars, however, these experiences will not come cheap. For those whose pockets do not run quite so deep, Richard Branson's Virgin Galactic and Jeff Bezos' Blue Origins are aiming to offer shorter flights into space for somewhere in the region of a million dollars. With trips expected to begin in the second half of 2021, space tourism will become an increasingly common industry much sooner than we might expect.

While there is clearly growing optimism about the potential for tourism however, the real money in space exploration is believed to lie further afield in the solar system. The mining of asteroids or other solar system objects for their resources and for precious minerals is another area that has traditionally been the subject of science fiction fantasy, but where the possibilities are already opening up quickly.

Governments are also increasingly keen to encourage commercial activity in this area, in part to reduce the reliance of major industries on materials sourced from other countries and safeguard against dwindling reserves on Earth. In 2015, the US amended their space laws to enable private entities to claim ownership of resources extracted from asteroids or other objects in space. They were followed not long after by Luxembourg who are also investing to stimulate the industry.

There have also been a small cluster of missions to asteroids whose objectives have been scientific rather than commercial, but which are nevertheless demonstrating the kind of technology that may be used one day be used to carry out prospecting activity and extract resources for Earthly purposes.

For several years, NASA were even working on a programme to capture a large boulder from the surface of an asteroid and transfer it into a stable orbit around the Moon. Again, while such a proof of principle mission was intended to support the establishment of a lunar base and enable the manufacture of technology beyond the limits of Earth, its progress would surely also have been followed closely by those with a commercial interest.

It is not just asteroids that could be a target for mining. The Moon has reserves of precious minerals and metals that would be easier to access than those of distant asteroids. It has also been suggested that the loose layer of deposits that covers much of the Moon's surface - more elegantly also referred to as 'lunar regolith' - could be useful for construction materials.

All of which means that, with the increasing accessibility of space and empowerment of commercial companies, the gold rush is surely only a matter of time. There is a significant amount of infrastructure that would need to be put in place to make mining feasible, but many now believe that the Earth's first trillionaires will earn their wealth from mining asteroids and that the flood could start within the next few decades.

If all goes well, there is much to be excited about in this expansion of humankind's presence beyond Mother Earth. Much of the effort of getting into space is spent in getting a fully laden rocket with all its fuel and cargo off the ground and building up speed as it passes through the first few kilometres of our dense atmosphere. It must fight not only against the Earth's gravity and the friction of the air through which it passes, but as it approaches the speed of sound, pressure waves build up around it, providing even more resistance and exerting dynamic stresses on the body of the rocket itself.

The enormous amount of energy required, as well as the need for the rocket itself to be aerodynamic, severely constrains how large a rocket payload can be. As things stand therefore, if we wanted to build large structures in space, there is an enormous cost to getting all the materials, tools and manpower into orbit in the first place.

If we were able to source the basic materials from asteroids or from the Moon, however, we would no longer need to expend all that effort transporting it into space. This would be a real game changer. Spacecraft could be built in space that no longer needed to expend the majority of their energy escaping the Earth's gravitational pull and the drag of its atmosphere. The efficiency saving could be used to either increase the payload capacity of the spacecraft or accelerate spacecraft to much higher velocities, making much longer trips possible across the solar system and eventually perhaps even beyond.

There could even be benefits for life on Earth as well. With resources being mined in space, there would be less pressure to plunder terrestrial reserves. The scars we have inflicted on our home planet could over time be healed; the habitats we have destroyed could return, and there might even be less conflict over valuable natural reserves as a result. If we could generate energy in space and beam it back to the Earth, we could even perhaps remove the need for polluting forms of energy production altogether, wiping out our problem with greenhouse gas emissions while easily being able to meet the energy needs of our ever-growing population.

We would become citizens of space. We would begin spreading out between different homes. The threat of a catastrophic extinction wiping out the human race and the other species of Earth, perhaps as a result of a meteor strike, would begin to be mitigated. Over time, maybe we would reach even further, and end up living between the stars. It is a dream that has captured the minds of many in the small hours of the morning, under the quiet of a crystal-clear night sky.

The vision is hard to resist, but there are also good reasons for us to be cautious. If the history of humankind's exploration of the Earth teaches us anything, it is that with every new achievement, we have also left a legacy of damage and destruction. From the decimation of native Indian tribes when Europeans first reached the shores of Central and Southern America, to the ugly litter of empty oxygen tanks and frigid corpses on the frozen slopes of Everest, we have a tendency to undermine our brightest moments of achievement by pushing too fast, too hard and too soon.

There are already warning signs that a rapid expansion of space exploration activity could quickly become overstretched. For one thing, the laws and regulations that govern spacefaring are poorly prepared for the world that lies ahead. The five main treaties that cover the international laws governing space exploration were all agreed in the midst of the Cold War at the end of the 1960 and 70s. Their chief concern - and that of the overarching Outer Space Treaty (OST) of 1967 - is to ensure that spacefaring activities remain peaceful and are carried out for the benefit of all humankind, reflecting the context of the time in which they were written.

As long as the main actors in space exploration have been national agencies, the legislation has worked well, and has acted as an effective safeguard. It is now over half a century since they were written, however, and there are inevitably gaps in their coverage that need to be addressed. In particular, the treaties were never intended to cover the activities of private companies operating in space, and their interpretation in the current context is ambiguous.

These holes will only get bigger and more problematic with time. Rather than rewriting the basic treaties, most experts seem to favour dealing with the ambiguities as and when they become issues. There is a lot of sense in this, not least because the principles enshrined in the original legislation are not ones that should be reopened for negotiation, but the approach also carries its own risks. While space is the most extreme environment we have attempted to explore, there are ways in which it is also extremely fragile, where small changes can have significant consequences.

Take the example of an asteroid that is mined so exhaustively that its mass ends up being substantially reduced as a result. This will have consequences for its orbit, which would need to be understood properly if we are to be sure that the asteroid would not be diverted in such a way that it could endanger other objects in the solar system.

There is no provision in law that provides guidance on whose responsibility it would be to assess the likelihood of orbital variations; what counts as an acceptable level of risk; what solar system objects should be considered protected; what the consequences would be for breaking the rules, or how they could be policed and enforced.

The threats could be partially mitigated by screening candidate asteroids prior to beginning of mining, capping the amount of material that can be removed, and monitoring their trajectories once operations are complete. It is by no means clear however how forward-looking governments will be in identifying such risks and their mitigations rigorously, or how diligent private companies will be in implementing the rules when the potential for profit is so huge.

The existing problem of space junk demonstrates how easy it is for issues that should have been (and indeed were) anticipated to escalate. Space junk refers to huge amounts of debris that continues to accumulate in the near-Earth environment as a result of our activities in space. This debris ranges in size from small flecks of paint to tools dropped by astronauts during spacewalks and defunct artificial satellites, all of which are zip through space, uncontrolled, at speeds of several kilometres a second. The positions of over 20,000 artificial objects are currently being actively monitored, but there is no way to keep track of the entire population of debris in space, which, if you include the smallest objects, is estimated to number over one hundred million objects.

While the majority of these objects are tiny, the speed at which they are travelling means that the damage they can inflict on spacecraft is substantial. It is now standard practice when estimating the efficiency and lifetime of satellites to account for the degrading of solar panels as a result of their constant sandblasting by small items of space debris. Larger items clearly present even greater risks. In a very real way, travelling through space is like trying to navigate an artillery range in which hundreds of thousands of bullets are flying around. The difference is that the bullets are far higher powered than those shot from a regular gun, and you have lost track of who shot them or what direction they were moving in. As a result, spacecraft routinely orient themselves in specific ways during flight to minimise the risk of critical damage from space junk, and the ISS even has to proactively manoeuvre itself on its orbit to avoid potentially dangerous collisions.

Despite there being widespread awareness of the risks since the 1980s and earlier, the issue is quickly reaching crisis point. Some even argue that space flight could end up being prevented altogether for a period of time if the density of space debris passes a certain limit at which cascades of collisions become self-perpetuating.

A significant amount of funding is now being directed to missions that aim to help clear up the detritus we have left in space so far, but there is a definite sense that more could have been done earlier had we been slightly more measured in our race to make use of space.

Space junk is a big concern, but there are potentially even more important risks that also need to be addressed in the grand future of spaceflight. One of the biggest is how we will safeguard any life that exists beyond the Earth.

Most habitats in the solar system are considered hostile to life as we know it. The likelihood is, therefore, that if extra-terrestrial life does exist nearby, it will be a form highly adapted microbes, potentially the lone survivors of previous biological kingdoms that died out.

Still, the discovery of life elsewhere in whatever form would be a true landmark in history, and there is little doubt that we would want to protect and study it in detail. There have already been a number of false (or at least inconclusive) alerts. One of the difficulties of identifying life, however, is removing the possibility of biological contamination in the instruments that are used to detect it. Rigorous steps have always been taken, therefore, to ensure all probes sent to solar system bodies that we suspect might support life are completely sterilised.

Once humans begin travelling to Mars, however, the risk of contamination becomes much greater. Accidents happen. In 2019, an Israeli lunar lander called Beresheet became the first privately led mission to attempt to land on the Moon. Unfortunately, technical problems as it was descending to the surface led to it crashing. The accident became well known in part because Beresheet was carrying a small capsule of tiny creatures called tardigrades that had effectively been frozen in suspended animation, as well as samples of human DNA.

This was controversial not only because tardigrades are among the most resilient animals on Earth, capable of surviving in the most extreme of environments including that of space, but because of claims that the addition of the animals and genetic material had not been disclosed to the regulating bodies involved in the launch, who were required to approve its payload.

Even if it had been disclosed, it is not clear that anything would have been done to remove them, as space law does not prevent the transport of biological material to the Moon, which is considered entirely dead. The implications had this crash happened on Mars, however, would have been much more serious. Quite apart from confusing the search for life, we do not know what impact foreign contamination of a planet or moon may have on any lifeforms that exist on them. It is possible that Earthly contaminants would pose a serious threat to the survival of Martian organisms. Should we be taking the risk when just one reckless action or piece of bad luck could have devastating implications?

Similar principles apply to attempts to terraform Mars. Any attempts to alter the planetary environment has the potential to destroy the habitat of any relic microbes. Should we even be considering such steps without yet having a clear answer as to whether life still exists there?

Part of the problem is that we simply have not had the kinds of conversations as a society that help determine where the boundaries should lie in our exploration of space, or that anticipate the problems and compromises we may end up having to face. The idea of a couple of individual asteroids being hollowed out for the resources they contain may not at first seem troubling, but the perspective becomes somewhat different if we think of asteroids as fossils from the formation of the solar system. Should these ancient witnesses of the past be preserved rather than destroyed, for future generations as much as for our current study?

Similarly, the ambition of setting up a Moon base as a steppingstone to further adventures in space is undeniably exciting, but the enthusiasm may fade away quickly if the character of the Moon began to be significantly changed as a result. Many sites of interest are located on the Moon's far side, which always faces away from Earth, or at the poles, but there are also substantial resources that would be of interest on its nearside too. How would public opinion regarding lunar mining change if the Moon's familiar face began to be disfigured as a result?

While such concerns may seem far-fetched at this stage, there will be little incentive for companies to take into account public sentiment once the technology and infrastructure are in place and the opportunity to make so much money is there for the taking.

It is also hard to imagine that we will be able to continue avoiding conflict in space once it has become another arena for competition and wealth. The Outer Space Treaty has worked well so far in ensuring space exploration has remained peaceful and in preventing nations from claiming sovereignty over celestial bodies. In the wake of the recent laws allowing private companies to claim ownership of resources collected from space, however, there is already a sense that these principles could soon begin to be undermined.

A report from the Harvard-Smithsonian Centre for Astrophysics published at the end of 2020, for example, highlighted how the interest in mining opportunities on the Moon is likely to quickly outstrip the availability of feasible sites. The report's authors warn how this could end up leading to crowding and interference between different teams located at the same mining sites. If conflict emerges between, for example, American and Chinese mining firms based on the Moon, it will surely not be long before those tensions also find their way back to Earth. It seems inevitable that such a situation would result in bolder claims being made about territorial rights and calls for the underlying principles of the OST to be re-examined.

Similarly, while private companies or nations cannot claim ownership of objects in space, the distinction becomes academic if the global characteristics of an asteroid is changed as a result of mining by a single organisation, or if the nature of a planet is altered through attempts to make it more Earth-like.

Underlying each of these issues is an even more fundamental question about the extent to which we have the right to mine objects in space in the first place, or to dream of terraforming an alien environment. It is an easy dig but given the devastating impact humankind has had on the natural environment of Earth, there is ample reason to claim that other solar system objects represent unique natural environments that deserve protection from ham-fisted tampering and exploitation by humans.

This right is never going to be questioned by those who believe in the unstoppable progress of humankind, just as the pioneers of the Industrial Revolution never stopped to consider whether the world they were building would turn out to be a sustainable one or not. We need such people to challenge and encourage us to be braver, or more inventive, or to dream bigger. We also however need the cynics and those who argue for checks and balances, to make sure we do not end up in a future where, once the excitement of discovery and the first flush of profit has passed, we discover that we have lost more than we have gained. At the moment, these voices seem to be missing from the conversation.

Once again, these anxieties may seem exaggerated given that the impact of our exploration on other worlds so far has been low, but if a brave new era for the space industry really is only just around the corner, the likelihood is that we will only begin asking ourselves what it is we value about space after irreversible changes have already taken place.

In a future when humanity's children play under the vaulted domes of the capital city of Mars; where the Martian atmosphere has been artificially thickened, and its ice caps melted to produce vast oceans and a weather system not unlike Earth's, will we celebrate our species' next major migration into space, or will we secretly regret the lost ruby-red glow of the Mars of old? We need to be preparing for this future now, for once the door is flung open and we embark on the path to this new adventure, there will be no way back again.

Printed in Great Britain
by Amazon